COMPUTER-AIDED DESIGN

Electronics, Comparative Advantage and Development

Computer-Aided Design

Electronics, Comparative Advantage and Development

Raphael Kaplinsky

A UNIDO Study
United Nations Industrial Development Organization

Macmillan Publishing Co., Inc.
New York

Macmillan Publishing Co., Inc.,
866 Third Avenue, New York, N.Y. 10022

Collier Macmillan Canada, Ltd

Library of Congress Catalog Card Number 82-14816

ISBN 0-02-949520-2

Library of Congress Cataloging in Publication Data

Kaplinsky, Raphael.
Computer-aided design.

Bibliography: p.136
1. Underdeveloped areas – Electronic industries.
2. Engineering design – Data processing. I. Title.
TK7836.K35 1982 338.4'762000425'091724 82-14816

Typeset by Donald Typesetting
Printed in the United States of America

CONTENTS

TABLE CAPTIONS

FIGURE CAPTIONS

ACKNOWLEDGEMENTS

This study could not have been completed without the co-operation of a great many people. Hermann Muegge of the Division of Conceptual and Industrial Studies of the United Nations Industrial Development Organisation provided a source of support where flexibility was an important element in allowing the project to be completed in the allotted time. A great number of people in the United Kingdom and United States computer-aided design supplying and using industries gave generously of their time despite my (slowly receding) ignorance; Al Gregory was particularly helpful in this respect. Various people – especially Erik Arnold, Norman Clarke, Kurt Hoffman, Mike Hubbard, John Marsden and Juan Rada – read and commented constructively on the first draft. Karen Brewer displayed her customary patience in coping with my missing of various deadlines. And finally, my family bore, as usual, the brunt of the ups and downs associated with the design of a project, the absences necessitated by fieldwork, and the struggle involved in writing up. My thanks to all of these named and unnamed individuals – the usual disclaimers apply.

PART I: THE SIGNIFICANCE OF ELECTRONICS

1 INTRODUCTION

Developments in the global economy over the past thirty-five years since the end of the war can be viewed in a variety of contrasting ways. At one extreme, little appears to have changed. The mass of the world's population continues to subsist at, or near, the bread line, and most of these people live in less-developed countries (LDCs). LDCs continue to rely on developed country (DC) technology and the flow of trade is predominantly one of DCs exchanging manufactured goods and technology for LDC primary products. But a contrasting – perhaps less pessimistic – view is also possible. Granted, LDCs continue as a whole to rely on DCs for technology, and LDCs continue to export predominantly primary products and import manufactures and technology. Yet within these overall trends, important changes have been taking place in the international division of labour. The share of LDCs in global manufacturing value-added has grown from around 7 per cent in 1965 to nearly 10 per cent in 1979 (UNIDO, 1980). And, no longer do LDCs continue to export primary products alone, for as the years have progressed so have their manufactured exports. By 1975 over 40 LDCs each exported over $100 million of manufactures, while aggregate LDC manufactured exports increased from $4.6 billion in 1965 to $55 billion in 1977 (Hoffman and Rush, 1980).

This growing participation of LDCs in global industrial output and trade reflects two inter-related tendencies in the world economy. The first concerns the specific attempts made over the years by newly independent LDCs to foster industrial production, which was seen as the source of future wealth. This involved the protection and subsidization of production, initially largely for local consumption and then subsequently destined for global markets. And the second factor fostering LDC industrialization was the growing presence of TNCs, actively seeking new markets in the fact of fiercely competitive oligopolistic markets (Knickerbocker, 1973).

Associated with these changes has been a growing capability by LDCs in the realm of science and technology. In the early post-war period educational facilities in almost all LDCs were minimal and an expansion of the numbers of educated, and an improvement in their quality, became a priority for most of these countries. This growing presence of skilled manpower in LDCs fostered and facilitated the transfer of technology. So (despite the inherent difficulty in measuring these phenomena), it seems as though the technological gap between DCs and LDCs began to narrow over the years. One indication of this has been that LDC manufactured exports have progressed beyond the 'mature' labour-intensive traditional industries (*à la* Vernon, 1966) and have increasingly come to encompass a variety of technology intensive goods (Lall, 1979; Katz 1978; O'Brien, 1981).

As a consequence of these developments there has been optimism that LDCs will be able to maintain their growing presence in the international divison of labour in industry – the 1978 UNIDO Lima Conference, for example, set a target for LDCs of 25 per cent of global value added in manufacturers by AD 2000 (UNIDO 1979).[1] But this horizon has in recent years become clouded both by the persistent and worsening recession in the world economy and by the rapid diffusion of efficient electronics-related technologies in DC manufacturing enterprises. These technologies not only tend to save labour (supposedly the source of LDC comparative advantage), but also provide other substantial benefits to innovating enterprises. So their differential diffusion in the world economy is likely to significantly affect the ability of non-innovating enterprises to compete in global markets. If this differential diffusion takes the lines of a DC–LDC split, then the technological gap, which appears to have narrowed in recent decades, may once again widen. In which case, the anticipated role of LDCs in the global division of labour in industry is likely to be less favourable (from the LDC point of view) than current perspectives suggest.

So far, however, all this is conjectural. Many observers, including those writing to the specific brief of DC governments (e.g., Nora, 1979; and Barron and Curnow, 1979), believe that electronics related innovations provide very substantial benefits to innovating enterprises. But little is known of the extent and nature of these benefits. Other observers (Rada, 1979; Kaplinsky, 1981) have argued that there is likely to be a differential diffusion of these technologies through the global economy, but once again little information has been available to back-up these assertions. It has become imperative, therefore, for more substantive evidence to be provided which might illuminate these generalized assertions.

It is in this rather striking context of our ignorance of the detailed impact of electronics technologies that this case-study on computer-aided design (henceforth referred to as CAD) should be seen. Although this sectoral study comprises an analysis of the specific origins, developments and uses of CAD technology as a sector in its own right, it more importantly illuminates general aspects of the origins and diffusion of electronics technologies, the benefits which arise from their use and the effect which they will have on the international division of labour. We believe that it is not possible to understand either of these issues (i.e., both the sectoral and the global trends) in isolation. But readers who prefer to confine themselves to the specific characteristics of the CAD sector might wish to limit their reading to Part II (chapters 3 to 8) and those with a wider interest in the impact of electronics on the global division of labour will find Parts I and III more to their interest. Essentially, however, the detailed analysis on the CAD sector provides incidental evidence to further discussion on the impact of electronics technologies on LDCs.

Before we proceed to the wider discussion on the relevance of electronics to the international division of labour (Chapter 2) it is necessary to explain briefly why the CAD sector has been chosen to illuminate these issues. This sector is peculiarly well-suited since it involves consideration of many of the more relevant issues. Thus, because of its downstream links with computer-aided manufacture (CAM) and information management, optimal use of CAD leads to *systems gains* in productivity. These systems gains – which arise from organizational re-structuring and optimized information management – lie at the heart of the 'electronics revolution', leading to the convergence of hitherto disparate processes such as communications and information processing and spawning a new family of jargon (e.g. 'telematics', 'informatics') and a restructuring of major TNCs (e.g. IBM and Xerox). The second advantage arising from a study of the CAD sector is that it illuminates the origins and development of *high-technology electronics sectors* which have, hitherto, almost all emerged out of the links between the United States defence, aerospace and electronics industries, initially in the form of small, independent firms and latterly as emerging TNCs themselves or as part of existing (not necessarily American) TNCs. And, thirdly, CAD technology is an excellent example of the *downstream diffusion* of electronics technologies to manufacturing industries where it is used to counter competitive pressures in the global market, partly arising from the growing exports of newly industrializing countries (NICs).

In order to explore both these particular facets of the CAD industry and the wider implications of electronics technologies in general, this study is separated into three sections:

(a) Part I is concerned with situating the development and diffusion of electronics technologies in a broader sweep of history. Electronics is considered in relation to long-run cycles of economic activity, and the discussion concludes with an assessment of the potential role which electronics technologies will have on the international division of labour in manufacturing. The major point made in this section is that electronics-related innovations are not merely one of a series of minor, unrelated technological developments. They represent a major set of new technologies which are of particular relevance to LDCs in their quest to become significant industrial producers.

(b) Part II is concerned with a detailed analysis of the CAD sector. In Chapters 3, 4 and 5 the origins, development and emerging market structure of the CAD sector are explored, together with an assessment of the significance of CAD for the wider diffusion of automation in manufacturing industries. This analysis is based upon field visits to suppliers, users and observers of the CAD industry in the United States and Europe.[2] Chapter 6 is based upon a series of interviews conducted with CAD users in the United States and Europe, and comprises an investigation of the benefits arising from the use of CAD. Of particular interest to LDCs is the question of whether CAD should be evaluated as a pure choice of technique capital/labour substitution decision (and if so, at what factor prices it becomes optimal), or whether it enables designers to achieve results which are not possible with manual design/draughting systems. The rate and direction of the downstream diffusion of CAD technology is then considered in Chapter 7, in which particular attention is paid to the industrial sector, and the types of benefits reaped by innovating enterprises in different branches, particularly those of relevance to LDCs. Then, given the imperative, if any, to the use of CAD technology, Chapter 8 assesses the skills required and the nature of the learning curves involved. The analysis is based upon evidence gathered in field visits to users and suppliers of CAD technology.

(c) Part III concludes the study with a discussion of the specific sectoral and more general policy implications for LDCs of emerging electronics technologies. It concludes with an assessment of the wider impact which electronics related innovations are likely to have on the international division of labour in industry.

Care is required in reading this study. As already pointed out, the analysis is ultimately aimed at exploring the likely impact of electronics related innovations on the international division of labour in industry. This is the primary subject matter of Parts I and III. But in order to fully assimilate the lessons to be learned from this sectoral case study, and to draw the appropriate conclusions, it is necessary to analyse the CAD industry in what might appear to some to be unnecessary detail – this is done in Part II. The reader is reminded that a selective reading may be appropriate – those interested in the global location of industry may wish to only skim Part II, while those interested primarily in CAD itself or in the electronics sector, will prefer to concentrate more attention on this section.

Notes

1 However this optimistic scenario was 'softened' at the subsequent UNIDO Conference held in New Delhi in January 1980 to a goal of around 15 per cent. This re-evaluation of the goal for AD 2000 was reached without recognition of the impact which electronics related innocations would have on the international division of labour in industry, a problem which is considered in this study.

2 The methodology, sample selection and sample characteristics are described in Appendix I.

2 LONG-WAVE CYCLES, ELECTRONICS AND LDCs

In this chapter we will be concerned to describe the fundamental importance of electronics in contemporary economic history. We will also explain why we believe that the diffusion of electronics-related innovations will be associated with changes in the international division of labour in industrial products. This discussion is a necessary prelude to the analysis of the CAD sector that follows, since it illuminates the context in which CAD is diffusing downstream to the manufacturing and allied sectors. It also illustrates how apposite is the choice of this sector as an example of electronics related innovations.

2.1 Long-wave cycles and electronics

Following two post-war decades of sustained and widespread economic growth, the worlds' economies are now, both individually and collectively, in a state of 'crisis'. It began with the LDCs in the early 1960s, spreading to some of the less dynamic DCs (the United Kingdom, the United States and Italy) in the 1970s and now even threatens the sustained growth of stronger economies such as West Germany and Japan. Not only are individual economies facing difficulties, but major multi-national systems of economic and political co-ordination stand on the precipice of disaster. Perhaps the best example of this is to be found in the international banking system which, increasingly reliant on debts incurred by vulnerable economies such as Poland, Zaire and Brazil, is repeatedly forced to co-ordinate rescue operations with both political and economic undertones.

Unlike the previous crisis of the 1930s, which was characterized by depressed demand, recession and stable prices, the current situation is one of stagflation, that is, simultaneous inflation and recession. Two differing explanations of this contemporary crisis are dominant in Western economic thought, namely the Keynesians and the monetarists. The former concentrate on the recessionary aspects of the contemporary situation, aiming to re-stimulate growth through demand expansion and hoping to contain inflation via reduced unit costs arising from scale economies and incomes policies of an explicit or implicit sort. By contrast the monetarists put great emphasis on the need to combat the inflationary aspects of the contemporary situation, arguing that the expansionary momentum of the Western economic system has been blunted by the ravages of inflation and the obtrusion of the demand-managing state in the sphere of accumulation.

Increasingly, however, a more historical set of explanations has begun to surface, situating contemporary crisis in a broader sweep of history.[1] Building upon a history of economic thought encompassing the writings amongst others of Marx, Schumpeter and a Russian writing in the 1920s called Kondratieff, attention has been placed on so-called long waves of economic activity, often called 'Kondratieff waves'.[2] Basically, the argument is that there are long wave cycles of economic activity and that contemporary stagflation must be seen as part of a downswing of the most recent cycle. While in the past these cycles appear to have had a duration of around fifty years, it is readily acknowledged that there is no justification for any fixed periodicity.

Great difficulties emerge in the measurement of these long-run cycles due to the inadequacy of historical data for almost all economies. Indeed a variety of attempts are currently being made in both European and North American universities attempting to provide flesh to this analytical skeleton. But, as Kleinknecht (1980) in his survey of various long-range theories argues, 'Although proof of long Kondratieff-*swings* thus appears very doubtful, one cannot ignore the fact that the data do reflect long-term fluctuations in the rhythm of growth' (p. 9). In support of this conclusion, Kleinknecht provides a summary of the various attempts made to measure the duration and intensity of these cycles, which is shown in Figure 2.1.[3] He cautions us that 'It must be kept in mind with regard to Chart 1, that the various authors present not only different explanations of long fluctuations, but they also have differing conceptions of the fluctuation patterns themselves' (p. 10).

But whatever the caveats with regard to the precise intensity and periodicity of these cycles, for the purposes of this study it is sufficient to note that there is now a fairly strong consensus that such long-wave cycles do indeed exist, that they have had a periodicity of approximately fifty years, and that – (with the exception of Rostow (1978) and Mandel (1980) – it is generally agreed that we do not yet appear to have reached the low point in the downswing of the most recent cycle.

The recognition of these cycles does not on its own remove the rationale for Keynesian or monetarist responses to stagflation, but it does of necessity place them in a different perspective. Most importantly as we shall see, it forces attention to the sphere of accumulation and the significance of technical change. And here the value of the earlier works of Marx and Schumpeter are most apparent.

In this context, Freeman (1978, 1979) and Clark, Freeman and Soete (1980), provide a particularly believable explanation for these long-run cycles. Their argument runs as follows.[4] Economic growth in the West has been fuelled by a supply-side motor, with entrepreneurs pursuing the goal of monopoly profit and achieving these profits through the innovation of new technological developments. Their achievements were reflected in a series of investments by competitors who were attracted by these supraprofits. The result was that competition increased and the monopoly profits were gradually whittled away.

But, as Clark *et al.* point out, even if this should indeed be the motor of economic growth, there is no reason why this in itself should lead to cycles of activity. The *cycles*, they argue, are in part created by inflexibilities, lags and imperfections in the behaviour of

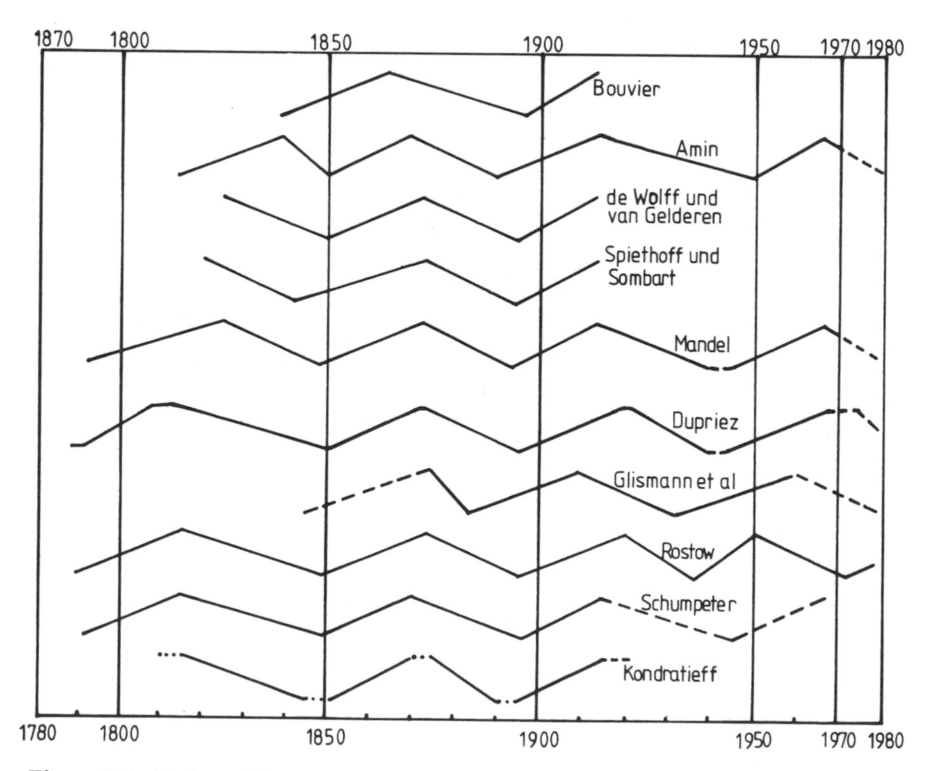

Figure 2.1 Dating of long-term fluctuations by different authors. (Source: Kleinknecht, 1980)

both capitalists and labour, and by the indivisibilities of fixed capital. Yet, despite the existence of these imperfections and indivisibilities, this does not necessarily imply the existence of cycles, since a sufficient number of randomly distributed minicycles should even-out any long-term fluctuations. So, they argue:

> Big wave effects could arise [only] if some of these innovations were very large and with a long time span in their own right (e.g. railways) and/or if some of them were interdependent and inter-connected for technological and social reasons. (Clark *et al.*, 1980, p. 25)

Freeman (1977), basing himself on the earlier pioneering work of Konratieff and Schumpeter, argues that over the course of the last 200 years, there have indeed been a series of major, '*heartland technologies*' which have fuelled these long-run cycles ('big-wave effects'). The first of these, beginning in the late eighteenth century was based upon textiles and the diffusion of the steam-engine; the second, with its onset in the mid-nineteenth century, was fuelled by the combined expansion of railroads and the diffusion of steel; the third, with its origins at the turn of the twentieth century, was based upon the internal combustion engine, electricity and the chemical industry; and the fourth, argues Freeman (1977), has been fuelled by electronics technology, beginning with the use of the valve in the 1930s, and proceeding with the invention of the transistor in the late 1940s, the integrated circuit in 1959 and the micro-processor in 1971.

Now in each of these heartland-technology based cycles, there is an 'expansionary' upswing and a 'rationalizing' downswing.

> In the major boom periods [i.e. the upswing] new technological systems tend to generate a great deal of employment, as the form which expansion takes is the installation of completely new capacity and since the technology is still in a relatively fluid state the new factories and plants are often fairly labour-intensive. New small firms may also play an important role among the new entrants and they tend to have a lower than average capital intensity.
>
> However, as the new technology matures [i.e. the downswing] several factors are inter-acting to reduce the employment generated per unit of investment . . . . Economies of scale begin to be important and these work in combination with technical changes associated with increasing standardisation. A process of

concentration tends to occur and competition forces increasing attention to the problem of cost-reducing technical change. (Clark *et al.*, 1980, p. 27)

So, whatever the specific difficulties associated with measuring the intensity and periodicity of the cycles, or the particular analytical motor provided in explanation, we cannot fail to recognize their existence and significance. But there clearly is a danger in being too mechanistic in any such interpretation, for each cycle has its particular characteristics. Thus, accepting (as Freeman and others argue) that electronics is the heartland technology of the contemporary cycle, we should still recognize that there are specific characteristics to it. By so doing, we can attempt to situate the current global crisis referred to in the first paragraphs of this report, as well as to analyse the particular role played by LDCs in the international division of labour in industry. [5] It also will help us, as we shall argue below, in providing a context in which this sectoral study on CAD is to be situated.

The first twenty-five years of post-World War II economic growth reflected a combination of two sets of factors. The first of these was the reconstruction that followed the devastation of war. The second concerned the expansion associated with the introduction of new products based upon the heartland electronics technology. Such products can be divided into four major categories (Freeman, 1981), namely consumer electronics (radio, television, record-players, tape-recorders, etc.), electronic capital goods (radar, computers, communication equipment, process control etc.), electronic components (valves, transistors, integrated circuits, resistors, etc.), and military equipment (radar, missile control systems etc.). In each sub-sector the introduction of new products was associated with rapid expansion of production, employment and trade.

By the end of the 1960s, reconstruction was largely complete, markets had begun to be saturated, [6] and as Mandel (1980) argues, there was also a systematic tendency towards under-consumption. Conjuncturally, the electronic cycle was moving into the downswing as major new product developments had been exhausted and more imitative firms had moved in to the sector, attracted by high profits and the rapid growth of markets. Thus not only were most markets characterized by increasing competition (which took national forms, with Japanese enterprises rapidly coming to dominate the consumer electronics markets and American firms doing well in other sectors), but the power of organized labour had begun to grow after a period of sustained near full employment, and changes in technology were forcing the investment of more capital intensive technologies. The consequence was that the 1970s saw a significant decline in the rate of profit in almost all economies (Hill, 1979); growing over-capacity in most major markets such as steel, shipbuilding and cars (OECD, 1979; Mandel, 1980); growing unemployment (Kaplinsky, 1981); inflation, and low/negative rates of economic growth. In the latter period further fuel was added to the flames of stagflation by the increase in energy costs, which reflected technological rigidities as consuming nations were unable to adjust rapidly to the changing relative price of energy. Next to the increase in energy prices, perhaps the most important development for the LDCs in the latter period was the decline in the rate of growth of world trade, which had expanded rapidly when the advanced economies were experiencing rapid economic growth and near-full employment, but fell off rapidly as these conditions no longer pertained.

2.2 LDCs in the context of long-wave cycles [7]

The same post-war period which saw the upswing of the electronics-based cycles also saw the decolonization of much of Asia and Africa and the growth of import substituting industrialization in both these continents and in Latin America. These industrialization strategies were based upon the restriction of imports with the concomitant growth of local production and a growing squeeze on foreign investment. Little attention was given in this period to the possibility of industrializing through the growth of exports since it was recognized that production in LDCs was unlikely to be competitive with DCs. The rationale for industrialization through import substitution was a growth in learning-by-doing and reduced unit costs as scale of production built up to optimal levels. It was only after these hurdles were surmounted that the comparative advantage of cheap labour could allow for competitively priced exports, or so it was believed.

The problem with these strategies was that once the easy early stages of import substitution were exhausted, the rate of economic growth began to slacken. This was in part because, characteristically, the productivity of capital in the intermediate and capital goods sectors was much lower than that in the consumer goods, 'final assembly' sectors. At the same time it became apparent that import substitution was proving to be an 'inefficient' way of conserving foreign exchange, and by the early 1960s many of the LDCs were beginning to experience severe balance of payments problems. However by the mid-1960s a new phenomenon was beginning to emerge in the world economy – a number of LDCs had either decided to change, or were induced to change their attitude towards foreign investment. Investment controls were relaxed and favourable tax and duty concessions were introduced to attract foreign investment which would produce for export. In addition favourable incentives were offered to indigenous firms to encourage them to export (Bergsman, 1979).

The results of this shift in policies were remarkable for a selected number of countries, leading to hitherto unheard of levels of growth in both exports and GDP. Korea's manufactured exports, for example, grew at an annual rate of 36 per cent between 1965 and 1975, with its *per capita* GNP growing at 7.3 per cent per annum in the same period. In the early phase these exports of manufactures were of the classic low-technology 'mature' products where LDCs competed on the basis of low wage costs (see Vernon, 1966). But increasingly, the technological gap between some LDCs and DCs began to diminish and LDCs came to be significant exporters of skill-intensive goods and services (see Plesch, 1978; Lall 1979; Katz, 1978, O'Brien, 1981). Despite the fact that only a limited number of countries – ten LDCs, for example, exported 78 per cent of all LDC manufactured exports in 1973 – profited from this extraordinary phase of export-led growth, the demonstration effect was enormous, with an increasing number of LDCs rapidly dismantling import substituting strategies and controls over foreign investment and substituting export-oriented policies, often based upon free trade zones. The international community also responded to their success – noting a growth in the LDC share of world manufacturing value added from 6.9 per cent in 1960 to 8.6 per cent in 1975 (UNIDO, 1980), the UNIDO Lima Declaration called for an LDC share of global output to expand to 25 per cent by AD 2000.

However all this occurred in a period in which, as we have seen, the long wave was in an upswing. Three particular facets of this upswing defined a role for LDCs in the inter-national division of labour. The first was a response by TNCs (initially American and subsequently, but to a lesser extent by Japanese and European competitors) to growing competition in the world market, which, as we have seen, is characteristic of the mature phase of the upswing. One way of cutting costs was for firms to decompose the labour process and have the labour-intensive elements undertaken in export processing zones in low-wage economies.[8] The second aspect of the upswing which defined a role for the LDCs was the very heavy requirement for labour in the heartland technology itself. Manufacture of the silicon chip, when produced in sufficient numbers, was cheap, with almost insignificant marginal costs – by contrast the packaging of these chips in their plastic containers and the insertion of connecting wires was a costly item, with little difference between marginal and average unit costs. Consequently there was intense pressure to reduce these assembling costs (which were predominantly labour costs) and the logic pushed firms towards using low wage labour in LDCs. And thirdly, the high levels of employment and economic growth in most DCs meant that cheap wage goods – predominantly those with a high labour input [9] – could be imported from LDCs without adding significantly to the levels of unemployment in the DCs.

The confluence of these three sets of related, favourable factors explains the great success in the growth of LDC manufactured exports over the past fifteen years. It also explains why so many LDCs began to institute policies designed to emulate the success of these economies – between 1978 and 1980 for example, the number of free trade zones increased from about 220 to over 350, most of these being in LDCs (Frazier, 1981). But it is precisely in this regard that the relevance of long-wave cycles, and the important role played by the heartland electronics technology, assumes its importance. Each of the three sets of favourable factors that have underlain export-led growth by LDCs is threatened as the current cycle moves into the supra-competitive, rationalizing downswing. Thus the favourable market-entry conditions into DCs for LDC manufactured goods – whether arising from 'runaway' TNC investments or from indigenous LDC firms – has begun to be eroded. Even before the most recent phase when the heartland electronics technology has begun to diffuse downstream in a cost-reducing role (which partly takes the form of displacement of labour by electronics-related innovations), employment in manufacturing sectors in DCs had begun to fall (Kaplinsky, 1981). At the same time the potential role of the services sector in absorbing this labour (Bell, 1974) appears to be threatened by electronics-related innovations (Freeman, 1977). Consequently, as unemployment rates have begun to rise in DCs, so have protectionist barriers, beginning in the most labour-intensive sectors (e.g., garments) and now spreading to other consumer (e.g., cars and television) and intermediate goods (e.g., steel). Secondly, the downstream use of electronics in other sectors has begun to undermine the comparative advantage of LDC firms producing with traditional technology and low-wage labour. And, thirdly, developments within the electronics sector itself, such as automated insertion of integrated circuits onto printed circuit boards, the packaging of the circuits themselves and the reduction of the number of circuits in many products due to the development of more powerful very large scale integration (VLSI), have diminished the requirement for cheap labour.

The burden of the previous discussion, therefore, is that the world economy has, over the past five to ten years, reached a minor turning point, which has fundamental implications for the global location of industry. In the preceding phase there existed a clear opportunity for export-led industrial growth in LDCs. But, now, many of the underlying phenomena that facilitated this type of industrial growth have begun to change. The source of this change has been the move from the expansionary upswing of the electronics-fuelled long-wave cycle to the 'rationalizing' downswing and as the heartland electronics technology has begun to diffuse to firms in DCs, so the technological gap between DCs and LDCs (which appeared to close somewhat in the 1960s and 1970s) has begun to widen once again. The potential impact of these developments on LDC industrialization is manifest.

2.3 CAD in the context of long waves

As we noted in the introductory chapter, and as we shall see in subsequent discussion, CAD spans both swings of the long-run cycle. In the earlier phase – that is in the late 1960s and early 1970s – it was a 'new' product, largely being used within the expanding electronics sector as an essential component in the design and manufacture of integrated circuits and printed circuit boards. But, recently (in the 'rationalizing' downswing), it has begun to filter down to established manufacturing sub-sectors where innovating enterprises have used it to optimize designs, reduce costs and shorten lead times in the face of growing competition. While the origins and development of CAD during the upswing are of intrinsic interest, it is clearly its role in the rationalizing downswing that is of greatest relevance to the subject under discussion.

More particularly we noted earlier that discussions of the likely impact of electronics were of a largely assertive nature. Not only do few empirical investigations exist [10] but with two exceptions (Rada, 1979; Hoffman and Rush, 1981) none of the research in this area considers the specific implications of the 'microelectronics revolution' for LDCs. This sectoral study on CAD aims to provide the sort of information necessary to evaluate the potential impact of electronics on LDCs. It does so by considering the *benefits* arising from the use of the technology, the *pace* at which it is likely to diffuse, and to which *sectors*, the *skills* required to utilize it and the extent to which it is filtering through to *LDC* plants. Only once these issues (and others, such as the extent to which TNCs' location decisions will be affected by these developments) have been clarified will it be possible to assess the extent to which LDC industrial participation in the global economy will be affected in the last quarter of the twentieth century by electronics related innovations.

Notes

1 There are also, amongst many other variants, monetarist explanations of long wave cycles. However, we distinguish here between two sets of analyses, namely those postulating short-run cycles (in which we include the rump of monetarists and Keynesians) and those arguing for the existence of long-run cycles.

2 So named after his pioneering work (see Kondratieff, 1935).

3 Unfortunately Kleinknecht omits the scale on the vertical axis, so it is not possible to gauge from his presentation the intensity of the amplitudes observed by the various researchers.

4 The following section is based largely, but not exclusively, on the collective works of these authors. An alternative and broader view on the relationship between long-run growth and technology sees the industrial revolution as having been fuelled by technologies which increased physical energy and dexterity. The coming 'second industrial revolution' argued to be based on electronics-related innovations, provides a family of technologies to enhance the processing of information, and hence intellectual activities (see, e.g., Rada 1979).

5 These ideas are treated in greater detail, and are related to the advance of automation technologies in general, in Kaplinsky (1981).

6 For example, by the late 1970s, there was one car on the road in the United States for every 1.2 licensed drivers, with 84 per cent of households owning at least one car. In Western Europe there was one car for every two adults. See Transatlantic Perspectives (1981).

7 Much of the discussion in this section is elaborated in Kaplinsky (1981).

8 But, as Sciberras (1979) shows, this occurred to the disadvantage of firms pursuing this path in the television sector. Although the American firms were able to cut costs by taking advantage of cheap labour in export processing zones, manually assembled sets were less reliable than those assembled in automated plants and the Japanese firms were able to capitalize on automated assembly, and dominate the American market.

9 Thus in the mid-1970s around 40 per cent of all LDC manufactured exports were in the shoe and leather and garment and textiles industries (Chenery and Keesing, (1978).

10 That is outside of corporate research establishments. For example the West German government paid Macintosh Consultants 1.4 million DM for a study of the likely diffusion of electronics.

PART II

COMPUTER-AIDED DESIGN:
AN EXAMPLE OF ELECTRONICS-RELATED INNOVATIONS

3 COMPUTER-AIDED DESIGN: THE TECHNOLOGY[1]

In this chapter we begin with a discussion of the place of design in production, pointing to the crucial role which it plays in defining the data base that is necessary for production and distribution. We then describe the interactive nature of computer-graphic systems, their historical evolution, the current configurations that are available and price trends. We conclude the chapter with a brief profile of the major firms selling turnkey system as a preparatory discussion for the analysis which follows in Chapters 4, 5 and 6.

3.1 The nature and role of design [2]

In situating the role played by design in the economy, it is necessary to view the production of commodities (particularly durable ones) and services in the light of systems. Here we might lean on insights gleaned from systems engineering literature:

> Successful planning and design of large complex systems requires the 'systems approach'. The systems approach recognises that factoring out a part of a problem by neglecting the interactions among subsystems and elements increases significantly the probability that a solution to the design problem will not be found; it requires that the boundaries of the system be extended outward as far as is required to determine which interrelationships are significant to the design problem.
>
> A system to be useful must satisfy a need. However, designing a system to just meet the need is not usually sufficient. With few exceptions, the system must be able to meet the need over a specified period of time in order to justify the investment in time, money, and effort. Thus one must consider a system in a dynamic sense – the life cycle or so-called 'cradle to grave' viewpoint. The system life cycle may be said to originate in the perception of a need and to terminate when the system becomes obsolete. (Kline and Lifson, 1968, pp. 12–15)

Thus design must be seen as a sub-set of a variety of interrelated activities which include production, installation, operation, maintenance and modification. Unless we recognize both this

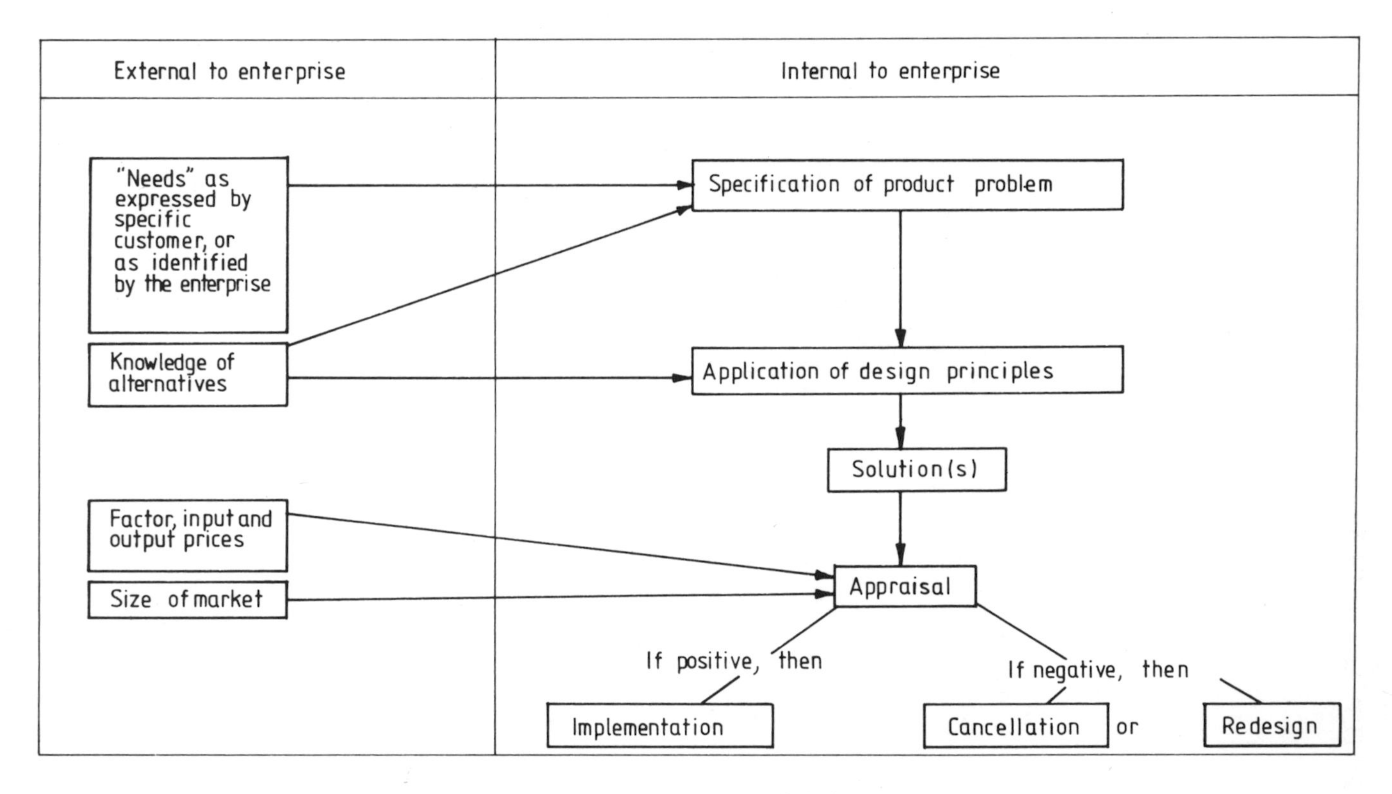

Figure 3.1 The nature of design

temporal dimension to product development and utilization, and the interrelationship between these different sub-sets of activities, we will be unable to fully appreciate the coming discussions on the benefits derived from using CAD technology.

In general terms, as shown in Figure 3.1 design comprises of four major sets of activities. Initially the product or design problem has to be *specified*. This can only be done by the enterprise if it has some view of the demand for the product – this may be very tightly defined (e.g., in a tender document) or loosely based upon a perceived 'hole' in the market. Clearly, underlying this specification is an awareness of the technologies available to execute the problem – not many designers, for example, would hazard the design of a motor car fuelled by water. Thereafter, in the light of practically available technologies, design involves the *application of design principles* to achieve the desired aims. This results in a unique *solution* or a series of alternatives, which then have to be evaluated in relation to factor, input and output prices to determine whether production will be profitable. It is also essential, at this stage, to consider the extent of the market since its scale and growth will have an important bearing upon the optimality of any of the available alternatives. In most cases the best-available alternative is sub-optimal and is subject to iterative *re-design*, in some cases involving a re-specification of the design problem.

In the most simple enterprises characteristic of pre-industrial revolution economies (and much of the Third World currently) where there was little division of labour, the design activity would not have been a specialized task. The same individuals who actually implemented the project would be involved in the design. But gradually as process and product technologies have become more complex, and as the market has widened to allow for the division of labour, so design has come to be separated from manufacture as an increasingly specialized activity.

While no end point in this evolving organization of manufacture was ever reached, this increasing specialization led to the development of a predominant form of organization, as represented in Figure 3.2 below. Essentially 'manufacture', (considered here to include a wide range of activities including mechanical, civil, electrical and structural engineering, publishing and communications) was characterized by three sub-divisions, namely design, production/distribution and information control. Each of these sub-divisions was specialized, situated in different buildings or areas and communicating predominately with each other and the outside world via various forms of paper-based activity. Most pertinently, the communications

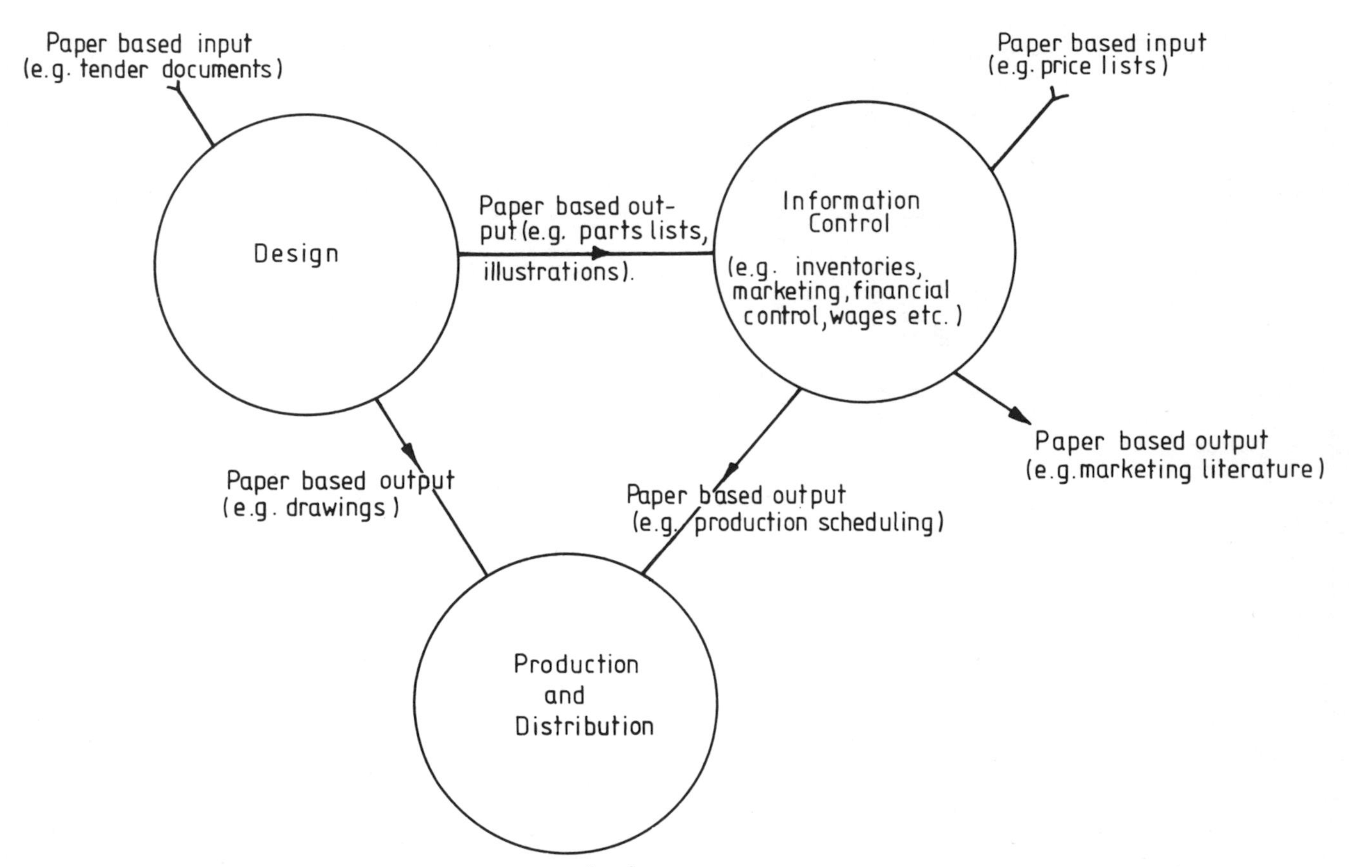

Figure 3.2 Pre-electronic organization of factory production

were predominantly from design to information control and from design to production/distribution rather than back to design. This information communicated from design was very often given in the paper-based form of drawings, a function performed by draughts-persons.

The use of computers (or more generally solid-state digital devices) from the mid-1950s, began to have an increasing impact [3] upon each of these three sub-divisions of manufacture. Their earliest use was in the design phase where the number-crunching capability of early mainframes allowed for the consideration of more complex design alternatives and gave a greater degree of assurance to the evaluation of design alternatives. This initially occurred in the defence-related aerospace sector but rapidly spread to other mechanical, civil, structural, electrical and engineering uses. The next major step occurred from the early 1960s when mainframe computers began to be used for information control in large enterprises, particularly in relation to payrolls and stockcontrol, and diffused to a wider number of users and smaller firms. The third phase of the diffusion of computing in manufacturing, from the mid-1960s onwards, saw the development of numerical controls (NC) initially in particular types of machine tools and subsequently spreading to production control.

Computer-aided design, or rather computer-*graphics*-aided design which is really what the CAD sector provides, is the most recent development in this electronification of manufacture. This technology has a number of characteristics that suggest that it is a key piece in the electronic jigsaw that will allow for the development of the automated factory. The first of these is that it is in the design phase that the data base that is internal to the firm (that is excluding prices or market conditions) is defined. CAD systems enable this data to be reduced to the basic binary building blocks [4] which are the currency of electronic systems in other parts of production and utilization. Consequently, automated manufacture is almost impossible without the prior electronification of design. Secondly, given the established work procedures in design offices this electronification of design could only diffuse widely when designers were given a graphical capability. For, as Besant (1980), argues:

> In most business or scientific applications a teletype, similar to a conventional typewriter, is suitable as an input and output device to a computer [which is a prerequisite for automation]. However, in engineering, particularly in design, the teletype on its own is not adequate as a communication device. The reason is that engineers traditionally communicate information graphically in the form of drawings. (p. 11)

The third characteristic of CAD that suggests it will play a key role in the move to the automated 'factory of the future' is its interactiveness. Traditionally the role of computers in design has occurred in a batch-processing environment where iterations made by designers were processed at some other time and often in a different location, thereby defining a highly specialized role for themselves. The ability that CAD has given to designers to interact iteratively in 'real time' with the model they are constructing has dramatically widened the scope for the electronification of design.[5]

The various elements of this electronic jigsaw are now mostly available. The state of the art lies in putting them in place. But to do so requires a new form of factory organization. In principle, and occasionally in practice, we can see this move to a new form of factory organization, which will be of as profound significance as the earlier move from the unitary, unspecialized pre-industrial enterprise to its three-sub-sector successor. The new phase of organization is represented in a comparison between Figure 3.2 and Figure 3.3 which illustrates a renewed merging of the three sub-sectors around a single data base as defined as the point of design. The consequence of this merging is a significant reduction in the need to communicate between the various sub-sectors, and where this occurs, for electronic communication to be substituted for paper-based systems. Perhaps more significantly, it involves a change in organization and a re-definition of the labour process – tasks which were formerly undertaken in production (e.g., parts programming, NC tape production) and information control (e.g.,

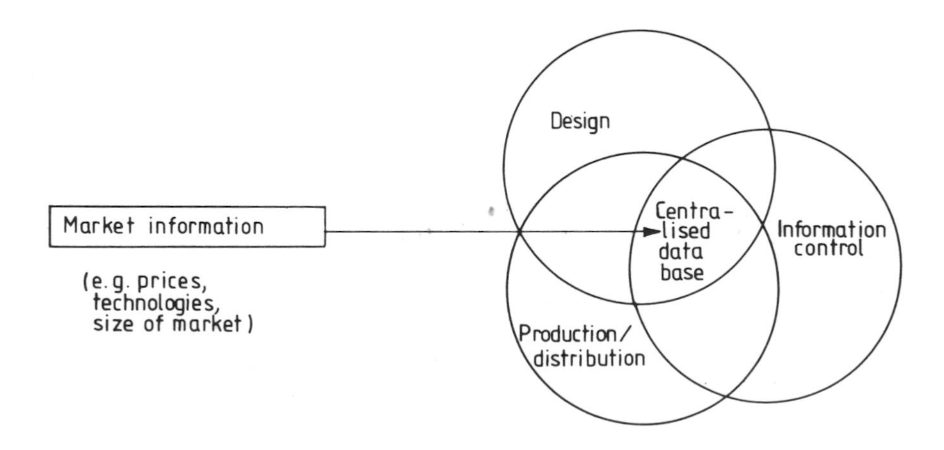

Figure 3.3 The move to the single system automated factory

parts lists, ordering, tender documents) can now be done in the design phase by utilizing a single unitary data base.

As we shall see, the technologies (particularly hardware) now exist to make it possible for these systems gains to be realized. The cutting edge of technical progress now concerns the integration of these various elements of electronification into a new system of production capitalizing on a unified data base. The major efficiency gain in the future will consequently be *systems gains*, based upon the re-organization of the factory, rather than on any productivity gains in a particular sub-activity. This change in emphasis in productivity growth will be reflected in subsequent discussion in Chapter 5 of the benefits arising from the use of CAD technology.

3.2 The nature of CAD

In essence a CAD system involves four different sets of hardware, as illustrated in Figure 3.4. Of primary importance is the computer, the central processing unit which:

> contains three main sections: the *controller* for sequentially examining each instruction and directing the action of the computer, the *arithmetic* unit for performing additions and subtractions which are fundamental to computers, and *core storage* where programs and data are stored for immediate use. (Besant, 1980, p. 21)

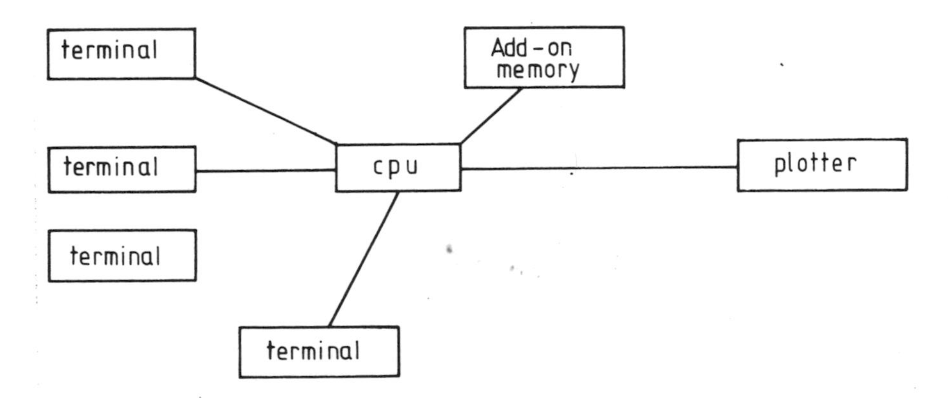

Figure 3.4 Typical CAD configuration

Data then has to be put into this computer, and here there are two major types of *workstation*; the first is the digitizer board, which converts designs to numerical co-ordinates (i.e., binary code), and the second is the visual display unit (i.e., a television-like screen) which allows the designer to see the design and to interactively proceed with or amend it. Once the design has been completed it has to be communicated – in paper-based systems this may take the form of drawings produced on various types of *plotting* devices, [6] while in automated systems, the design is transmitted in the form of numerical co-ordinates, often on paper tape. The final set of hardware is *add-on-memory*, since with current technology the viable computers do not possess sufficient core-memory of their own. Once again there are alternatives (basically disc and tape), each of which has its advantages and disadvantages and are often used together in the same CAD system.

On its own this hardware, all of which is readily available from a fairly large number of suppliers, is useless since it requires a series of operating instructions – that is, software – to operate. Basically there are two different elements of software that are necessary for draughting, with an additional capability for design. The first of these is the *operating system* which specifies in fact the way in which the computer executes its tasks – all computers require an operating system. But to produce drawings it is necessary to also have *basic graphics software*, which is the ability to construct lines, circles, arcs, rectangles, mirror images and so on to the screen and subsequently on to paper or paper tape. The software required for graphics is relatively simple and is now relatively widely available. But the use of the CAD system for design requires specific and more complex *applications software*. The earliest applications programs were developed for the electronics sector. Subsequently they began to meet the needs of the mechanical engineering sector and the future push appears to be occurring in the architecture, civil and structural engineering sectors. There are already a wide variety of different applications programs, the development of which (as we shall see) has been the heart of the activities of CAD suppliers. These range from by now fairly standard routines (e.g., the calculation of optimum routes for linking the electronic components on a printed circuit board), to those which are actively under development (e.g., solid modelling).[7] Some of these applications programs are not heavy users of data (e.g., auto-routing), whereas others (such as finite element stress analysis) require powerful mainframe computers for execution. From the operator's point of view the primary method of using these applications programs is via 'menus' which are removable sensitized tablets that contain specific routines relevant to particular applications programs and enable rapid use.[8] Characteristically an efficiently run CAD system will have very many menus, each designed to fit a particular set of applications.

The distinction between these three variants of software is, as we shall see, of critical importance to understanding issues such as the dynamics of market structure, the benefits arising from the use of CAD, the skills it requires and their associated learning curves.

3.3 Types of CAD systems

As we will see in the coming discussion of market structure, there are a fairly large number of firms supplying both packages of CAD systems and/or individual hardware and software components of such a system. Moreover most of these suppliers provide a wide range of options — Intergraph, for example (currently ranked fourth in sales) offers fifty alternative terminal configurations. But the most significant of these options arises from the type of computer that is used and this largely defines, as we shall see, the three segments of the CAD market. These three alternatives are:

(a) Microprocessor-driven dedicated terminals, which are small and not very powerful. Basically they are suitable as pure draughting aids — a sort of draughtsperson's word processor — although some are also able to undertake elementary processing programmes such as laying out the electrical circuits on a printed circuit board.
(b) Minicomputer-driven systems are more powerful and more flexible. These form the basic processing capability for all of the existing turnkey systems. Their strength relative to the small dedicated systems is that they are powerful enough to be able to undertake a large number of applications programs as well as to function as a draughting tool; each minicomputer is also able to drive between three and eight terminals, depending upon the particular suppliers' software and the use made of it by the user.
(c) The power of mainframe computers provides two major advantages to users. The first is that these systems are powerful enough to undertake the more taxing requirements of particular software applications (e.g., finite element modelling in mechanical engineering) as well as to also process data bases (e.g., parts lists, payrolls, etc.) for which minicomputers are not suitable. And second, the power of the mainframes allows large users (or those using them on a time-sharing basis) to reap economies of scale in unit terminal costs.

3.4 The price of CAD systems

The consequence of using these different types of computers is that there are economies of scale in purchasing any systems that run with minicomputers or mainframe computers. This is because unlike the microprocessor-driven systems (which have a separate computer for every workstation) each of the more powerful computers are able to support a number of terminals. Thus the minicomputers can cope with up to eight terminals (depending upon use), while some of the mainframe computers can support over fifty terminals with additional capability available for other batch-processing tasks (such as payrolls).

Given the wide variation in the processing capabilities of the various systems, and in the range of applications programs provided by each vendor, it does not make a great deal of sense to compare the prices of the alternative systems. But basically it is possible to obtain (see Figure 3.5) a single-terminal microprocessor-driven system – providing an elementary graphics capability with perhaps one or two simple applications programs – for around $30,000–50,000. Minicomputer-driven systems have greater applications capabilities – they are consequently more expensive (around

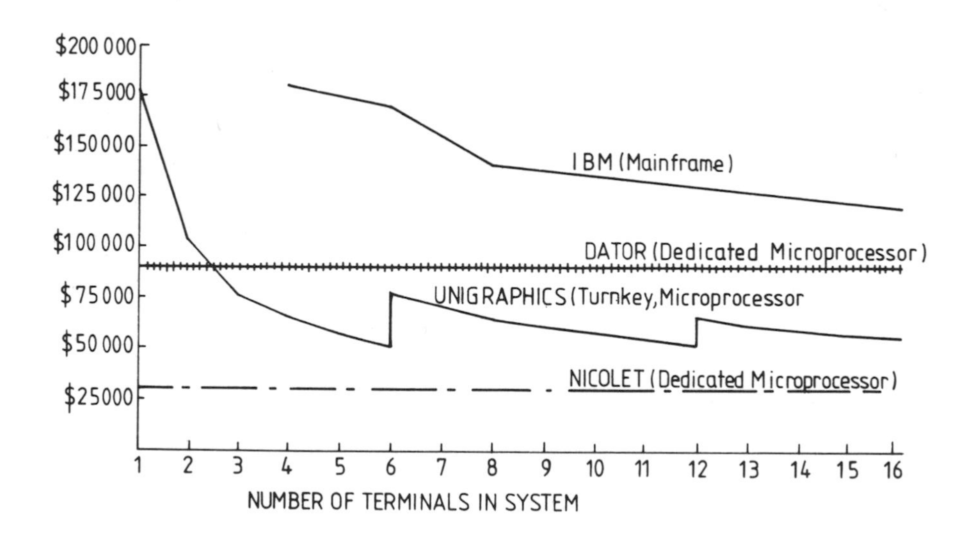

Figure 3.5 Unit terminal costs; some examples.
(Source: field interviews and Anderson Report, Vol. 2, No. 6)

$70,000 per terminal, with software) and show limited scale economies. The mainframe systems naturally provide far greater capabilities (although the vendors have not generally developed as comprehensive a suite of applications programs as the minicomputer-driven systems) with more significant economies of scale: but in general they tend to have the highest unit terminal costs.

However, whatever the variation in entry costs or unit terminal costs between these three different types of CAD systems, they all represent a considerably more significant outlay for the drawing office than the $2,000 per head worth of drawing equipment which backs draughtspersons in traditional DC drawing-offices.

Although much of the cost of CAD systems is comprised of software input and other overheads (which will be discussed at length in the following chapter), the remarkable decline in electronic hardware prices has led to a sustained drop in the real price of CAD systems. Thus the first system with a graphics capability was developed by IBM in the 1960s and was used by the aerospace industry – it cost $3 million (Calma, 1978). Compared with the average *system* price[9] of Computervision CAD equipment (which in current prices was around $394,000 in 1976, $403,000 in 1977, $345,000 in 1978, $412,000 in 1979 and $347,000 in 1980) [10] and assuming a broadly similar configuration of terminals per system over these five years, this represents a sharp decline in real prices (i.e., taking account of inflation) of CAD systems.

3.5 The origins of CAD

As with the earlier development of the semiconductor industry (Braun and MacDonald, 1978), the origins of the CAD sector are enmeshed in a protracted interaction between the military, the aerospace, the automobile and electronics sectors, spin-offs from academic institutions, and venture capital. Although little is known of the details of this interaction, it is nevertheless illuminating to retrace what is known since it provides insights into the global evolution and concentration of high technology industries.

The military/aerospace sector has played an important role over the years, particularly in the United States. Thus the first technological breakthrough for CAD – that is, the refresh graphics screen plus light-pen, which allowed for an interactive relationship between the screen and the operator – was developed for the SAGE early warning radar system in the 1950s. Later, in the 1960s, the United States Department of Defence played an important role in disseminating the virtues of CAD, culminating in a major conference between defence and industry in 1968 when the concept of CAD/CAM (computer-aided design/computer-aided manufacture) began to receive prominence. This concern with automation (i.e., CAD/CAM rather than CAD alone) remained a preoccupation of the military in the 1970s leading, via the work of the Air Force Integrated Computer-aided Manufacturing (ICAM) Program (funded jointly by the United States Air Force, the Army, the Navy and NASA), to the establishment in 1979 of a committee attempting to implement common software standards which will allow different CAD/CAM systems to intercommunicate with each other. The military also played a significant role in the origins of particular firms, especially Intergraph, which has been the most rapidly growing CAD vendor since 1976 and is now the fourth largest firm in the industry. Intergraph had its origins in writing large scale integration (LSI) software for the United States army; but most importantly it was able to recruit software specialists from the neighbouring NASA centre in Huntsville, Alabama after the winding down of the Apollo Moon Program. Not only did this occur during a period of an acute shortage of experienced software writers, but the geographical isolation of NASA provided a more stable cadre of software writers than any of Intergraph's competitors.

A second major stimulus to CAD technology arose from the non-military aerospace, automobile and electronic sectors. General Motors instigated the first industrial application of CAD with its design augmented by computers (DAC) program, beginning in 1959. This was followed shortly afterwards by similar programmes in Lockheed and McDonnel aircraft firms in the United States. The United Kingdom CAD industry's historic pre-eminence amongst European countries (especially in relation to software) is attributable to the relatively advanced state of its aircraft industry in the 1960s (Rader and Wingert, 1981). Over the years these three pioneering American firms provided a cadre of trained software writers who were instrumental in the formation of many of the major turnkey CAD vendors in the 1970s, as well as in the development of the basic graphics software which these firms used.

The inter-relation between the CAD and electronics sectors was vital to the development of both. For the CAD industry, the electronics sector provided a source of experienced manpower, with Auto-trol being rescued from a managerial crisis by ex-IBM personnel in 1974 and Intergraph being founded by ex-IBM employees. The electronics industry in turn benefitted from the availability of CAD technology (since modern LSI, VLSI and ULA chips *cannot* be designed with CAD systems) and provided a ready market for the new industry's output.

A third factor facilitating the emergence of the CAD industry was the computer science departments of particular universities. In the United States the maturing of software development in the Boston area universities led to an emigration of personnel from the Massachusetts Institute of Technology (MIT) who founded Applicon and Evans and Sutherland. In the United Kingdom a number of spin-offs from Cambridge and Leeds universities have led to specialized software firms who are at the forefront of the emerging solid-modelling software. Similar developments have occurred in Scandinavia (e.g., ICAN in Norway) and Germany and France.

The final key to the emergence of the CAD sector was finance. In the earliest period, when the low levels of software development enabled easy entry, capital was often provided by the entrepreneurs themselves (as was the case with almost all the major turnkey vendors). But the explosive growth of the industry required extensive working capital which in the case of Auto-trol and Applicon meant recourse to venture capital, and to takeovers by larger firms in the case of United Computing (purchased by McDonnel Douglas in 1976, who changed its name from United Computing to Unigraphics), and Calma (acquired by United Telecom in 1978 and sold to General Electric in 1981). Another source of finance was the stock market, and five firms – Computervision (various share placements have been made since it first went public in 1973), Auto-trol (raising $6.5 million in 1979 and $11 million in 1980), Applicon (selling 31 per cent of its equity in 1978), Intergraph (raising $27 million from the sale of 13.2 per cent of its equity in 1981) and Gerber (raising $7.1 million from the sale of 20 per cent of its equity in 1981) – offered part of their stock to the public. All of these turnkey vendors were forced into both long- and short-term debt to finance their expansion and, indeed, one of the major functions of the offering of shares to the public by the five public companies was to repay extensive accumulated debt. Thus prior to its public flotation, long-term debt of Intergraph, for example, jumped from $842,000 in 1979 to $6,231,000 in 1980 to finance growth in sales of 53 per cent over this period.

As with the semiconductor industry itself (Forester, 1978), the CAD industry was stimulated by the rapid turnover of manpower. The most significant of these in the early years was a group of specialized software writers who established themselves (as a firm called Systems Science Software) on the United States West Coast in the early 1970s, and whose activities were instrumental in the development of all of the existing minicomputer-based CAD suppliers. They provided, through a series of complex alliances with, in turn, Computervision, Gerber, MCS [11] and Calma, the basic software for mechanical applications for almost all the existing software packages now available. In the most recent (post-1978) period a new phenomenon has begun to emerge of small firms, often producing dedicated, microprocessor-driven systems, begun by ex-employees of the major turnkey vendors.[12]

Finally, a description of the origins of the CAD sector would be incomplete without a mention of the most significant developments in hardware. Aside from the general decline in memory and component prices (averaging 30 per cent per annum since 1959, see Noyce, 1977), the most important developments were:

(a) the interactive refresh + light pen graphics tube in the 1950s;
(b) the introduction of the storage tube (for screens) in 1970 by Tektronix;[13]
(c) the development of the minicomputer in the early 1970s and the microprocessor in the late 1970s; and
(d) the improvement in the most recent period of a television-type video screen (raster) which is quick to use, is cheap and allows for the incorporation of colour in applications software. Its major disadvantage at present is poor detail; but the technology continues to improve in this regard.

3.6 A profile of major turnkey suppliers

Before we proceed to the analysis of evolving CAD market structures in Chapter 5, it will be useful to set the scene by providing a short profile of the major suppliers of turnkey systems. (More detailed information for each of these firms is provided in Chapter 5.) Currently all of these are of American origin, although a number of European and Japanese firms are beginning to compete in the same sectors.

The dominant firm in the industry is *Computervision*, which was established in 1969 producing CAD systems as well as equipment for the manufacture of semiconductors. Over the decade the CAD division became increasingly dominant and the semiconductor division, Cobilt, was sold off in early 1981. A projected downstream integration into robot production proved to be divisive, leading to the departure in 1979 of one of the firm's founders, Paul Villers who established an independent robotics firm. [14]

Traditionally strong in the electronics sector, but moving rapidly into mechanical engineering and other fields is *Calma*. Established in 1964 this firm was largely dormant until the early 1970s. It was then purchased by a large telephone firm, United Telecoms, in 1978 and after a repeatedly increased offer, taken over by General Electric in 1980.

General Electric at that time already owned around 29 per cent of the equity of *Applicon*, a firm with traditional strength in mechanical engineering applications of CAD, and would have apparently preferred to enlarge this holding rather than to take over Calma. However the MIT graduates who founded Applicon in 1970 refused to relinquish control to GEC and were taken over subsequently by a French multinational corporation (originally specializing in petroleum-extraction equipment), Schlumberger, in late 1981.

Intergraph, based in Alabama near NASA headquarters, is the most rapidly growing CAD firm, and is currently trying to broaden its products out of the mapping sector and into mechanical and electronic applications. It delivered its first system in 1973 and has profitted enormously from its proximity to NASA (which provided a steady stream of software writers) and software contracts with the Department of Defence.

Auto-trol has traditionally been the other CAD supplier with an expertise in mapping, although it too has been moving into other applications areas in recent years. It was established in 1962 but remained small until it was taken over by venture capital in 1973.

Unigraphics, now the subsidiary of McDonnel Douglas Aircraft responsible for sales of turnkey CAD systems, has a long history. It was an early entrant, established as United Computing in 1961 to develop software for numerically controlled machine tools. This explains its strength in mechanical engineering applications, which was reinforced when it was taken over by McDonnel in 1976. It is interesting that McDonnel is also beginning to sell – in a largely independent operation – a mainframe-based CAD system called CADD (standing for computer-aided design and draughting).

The *Gerber Corporation* has a long history in plotting technology which was strengthened by its subsequent development of a computerized garment-cutting technology in which it is the world leader, having recently taken over its main rival, a subsidiary of Hughes Aircraft Corporation (see Hoffman and Rush, 1981). These strengths led it into CAD applications where it has developed a particular capability in relation to mechanical engineering applications, with considerable experience in the aerospace sector.

Finally, the only significantly sized mainframe-based supplier is *IBM*. For this very large TNC, CAD is currently a minute sector, representing around 0.003 per cent of turnover. Nevertheless, as we shall see in Chapter 5, the enormous size of IBM and its comprehensive range of capabilities, make it a very powerful competitor in the CAD sector, to which it began to give more serious consideration after 1974.

Notes

1. Readers who are not familiar with the specialized (and often rather peculiar) jargon of the computer industry, should consult the glossary at the end of the book.
2. The discussion in this part of the book, particularly with respect to the wider phenomenon of automation, is treated in much greater detail in Kaplinsky (1983).
3. The degree of diffusion and impact was influenced by the remarkable and sustained lowering of hardware costs (see Noyce, 1977).
4. The basic language of all electronics is information provided in an either/or, yes/no (conventionally 0 and 1) form. This is referred to as binary code. For a further discussion of this and other issues see Laurie (1980).
5. In the light of these observations it is clear that by CAD we really mean interactive computer-graphics-aided design. However in this study we follow the established procedure of distinguishing between this form of CAD and the more traditional form of non-graphics batch processing which we refer to more generally as 'the use of computers in design'.
6. The main variants of which are flat-bed plotters and drum-roll plotters (both of which are what they seem), hard copy devices (similar to xeroxing) and electrostatic printing (the most modern and expensive variant which produces the fastest and best quality copy).
7. Currently the three-dimensional (3D) applications software of almost all CAD suppliers is built-up of a series of wire frames. These, although requiring substantially fewer data inputs and processing capability than real 3D solid modelling packages, have a number of disadvantages associated with the fact that they do not differentiate between 'space' and 'real matter'. Consequently they provide significant hurdles to some volumetric analysis and are inhibiting the development of automated assembly techniques. It is anticipated (in 1982) that the development of efficient solid-modelling software, which is currently underway in all CAD suppliers and numerous universities and research institutes, will take another two to four years.
8. An analogy to changing menus would be the change from a statistical calculator (with buttons for regressions, square roots, etc.) to a scientific calculator (with buttons for trigonometric functions).
9. Note that this refers to system's costs. From the user's viewpoint, unit terminal costs would be more useful, but the relevant data is not available to make these calculations.
10. Derived by relating annual turnover to declared number of systems sold.
11. MCS (Manufacturing Consultancy Services) was founded and headed by Patrick Hanratty, a flamboyant character who led one of the original CAD teams in General Motors and was instrumental in the formation and early activities of Systems Science Software.
12. For example, a very recent entrant to the sector is the Graphics Technology Corporation. It was begun by Ronald McElhaney who spent two years as a software writer with MCS and was then recruited by Auto-trol (where he worked for only three months) to head their integration of mechanical software bought-in from MCS. McElhaney found $4 million of venture capital to assist him in establishing the firm (see *Harvard Newsletter*, Vol. 3, No. 7–8, 1981).
13. The storage tube had the major advantages over the refresh screen of not flickering (and therefore being easier to use) and lower price. Its major disadvantages are that it is slower to use (since every interaction requires the screen to be 'repainted') and requires subdued lighting.
14. In fact, Villers departure has led to a law suit. Villers had been delegated by the

Board to develop a feasibility study for this robotics project, which the Board subsequently deemed to be too ambitions. After his departure from Computervision Villers was sued because he had used company time to develop his new corporate plan. Despite Computervision's current renunciation of robot production, some observers feel that the logic of this vertical integration will force such a step on the firm in the near future.

4 THE MARKET FOR CAD EQUIPMENT

In this chapter we survey the size of the market for CAD technology and attempt to project its growth path over the coming decade. This is done in aggregate, as well as by sector and region. It is a necessary prelude to our later discussion on diffusion and the potential impact of CAD (and other electronics-based industries) on LDCs.

4.1 Market size and growth

Estimating the absolute size of the market for CAD systems is almost impossible since turnover data for suppliers only really exists for the major United States vendors. Collectively, these vendors had sales of around $575 million in 1980 (see Table 4.1) to which must be added the sales of European [1] and Japanese firms [2] as well as those of animation and business graphics systems, which are specialized sub-sectors of CAD. The business graphics sector alone in the United States is expected to be worth around $200 million in 1981 (*Harvard Newsletter* Vol. 3 No. 9, 1981). Additional systems of an unknown value have been installed within firms, often designed to meet custom needs – the most significant of these is probably IBM electronics CAD software which is not marketed externally. In the light of these observations it would be surprising if the value of global CAD systems was much less than $1 billion in 1980. This is a significantly sized market and compares favourably, for example, to the estimated value of colour television sales in the United States of $3.6 billion in 1980 (*Electronics International*, January 1981).

Table 4.1 assembles the available data on the growth of the major United States turnkey vendors. For most of these firms the figures on turnover are incomplete since only the publicly owned companies are obliged to publish balance sheets. But with the exception of IBM (where the estimate of turnover is based upon interviews) the data in this table is fairly accurate. Looking at their turnover since 1976 (from when fairly complete records are available for most firms) the industry's turnover growth has been explosive. Between 1976 and 1980 the annual compound growth rate was 69.3 per cent, rising to 84.6 per cent between 1978 and 1980. In Figure 4.1 (a) we can compare the growth of these total sales to what they would have been had the CAD industry grown at the same rate as the world's largest electronics firm, IBM. We also compare its growth to that of the most successful firm (Digital Equipment Corporation) in the minicomputer industry, a sector which is generally recognized to be one of the most dynamic industries of the 1970s. In both sets of comparisons the CAD industry emerges with a startlingly high relative rate of growth.

Of course many other new industries have shown such high rates of growth in their early, formative period. The major question for the CAD industry, therefore, is whether, as the industry matures, these high rates of growth can be maintained, and for how long. It is naturally impossible to provide an unequivocal answer to this question, but it is possible to draw some tentative conclusions. First, *all* of the firms visited in the course of this research had order backlogs [3] which suggested that the 1978–1980 growth rate would probably be equalled in the 1981–1982 period. Second, this order backlog occurred in the context of a global recession which has had a severe impact on both the electronics and other manufacturing industries, almost all of which are facing very low, or negative, growth rates. Third, the cost-reducing and quality-improving benefits arising from the use of CAD technology suggest that it is precisely during the recessionary downswing of the long-run cycle that the technology will be in greatest demand. Fourth, one major sector of the market, that is low-cost draughting systems, has only just been opened up.[4] And fifth, the market for non-electronic CAD equipment has barely been opened. In United States manufacturing, penetration is variously estimated at between 5 and 10 per cent of potential users, with penetration in other sectors such as civil engineering, structural engineering, architecture and mapping being even lower; in Europe the extent of penetration is even lower.

On the basis of these observations it is possible to make some estimates of the size of the CAD market covered by these US turnkey suppliers in the future, as can be seen from Table 4.2 and Figure 4.1 (b). By 1984 the market for these turnkey CAD systems is likely to approach $4 billion per year:[5] beyond 1984 it is impossible to predict, but if the market continued to expand after 1984 at a lower 20 per cent per annum, then the annual market for CAD equipment in 1990 looks like exceeding $12 billion. Considering some of the other CAD sales mentioned earlier, it would be surprising if combined global sales did not exceed $5–6 billion in 1984 and $15 billion in 1990. As a point of comparison the current annual global market for robots is around $350 million and this is expected to rise to only $2 billion by 1990 (*Financial Times*, 19 May 1981); while the projected annual sales of colour televisions in the

Table 4.1 Global sales of CAD/CAM turnkey systems by major United States vendors

	Date of origin	1969	1970	1971	1972	1973	1974	1975	1976	1977	1978	1979	1980
Applicon	1968								10,183	16,640	18,372	28,469	50,776
Auto-trol	1959			±1			1,447	4,835	6,971	12,549	21,850	33,540	51,000
Calma	1964							9,000	10,000	17,000	27,000	43,000	79,400
Computer-vision *	1969	51	724	2,567	5,118	8,510	13,342	14,572	19,647	28,188	43,432	103,004	191,000
Gerber	1974								2,500	1,400	4,400	10,200	23,406†
IBM ‡	1974								15,000	20,000	23,000	40,000	70,000
Intergraph §	1969	50	270	480	620	920	2,000	3,200	5,718	9,173	20,146	29,518	56,468
Unigraphics ¶	1961										5,500	7,500	13,000
Other **											5,000	21,000	40,000
Total ††									70,019	110,000	168,700	316,231	575,000

* Excludes Cobilt (i.e., CAD/CAM only).
† Includes estimate of 100 sales of PC800 terminals by Gerber Scientific.
‡ Estimate based upon interviews, not balance sheets.
§ Includes sales of non-graphic software to Government which was 6 per cent of turnover in 1980.
¶ Includes estimate for CADD, excludes sales of UNIAPT.
** Based on Kurlack (1980), excluding Unigraphics but making allowance for underrepresentation of small systems.
†† Excludes animation and graphics CAD. Graphics CAD alone worth around $200 m in USA in 1981.
Source: Annual Reports and Interviews.

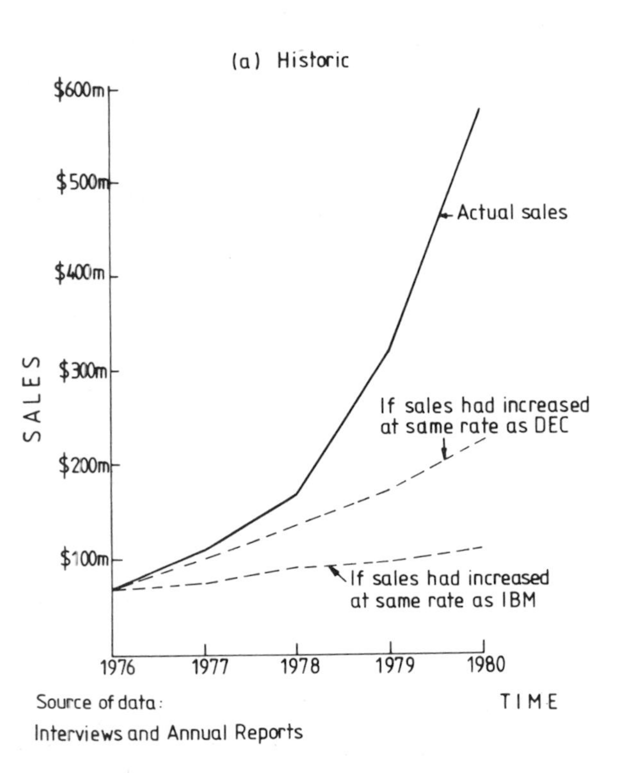

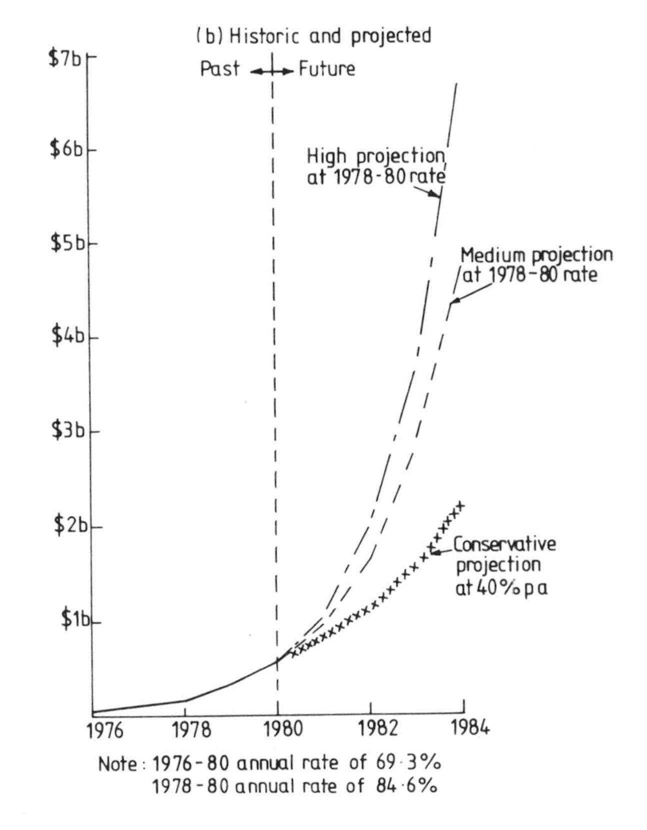

Figure 4.1 Sales of United States turnkey vendors – past and projected. (Source: interviews and annual reports)

United States in 1984 is only around $4 billion, and that for semiconductors is $12 billion (*Electronics International*, January 1981). By all counts, therefore, CAD is a very large sector of activity.

Table 4.2 Estimate of future sales of CAD equipment by major United States turnkey vendors, and for sales of colour televisions in the United States ($ million)

	1980	1984
Actual	575.05	
High estimate (1978–1980 rate of 84.6 per cent per annum)		6678
Medium estimate (1976–1980 rate of 69.3 per cent per annum)		4724
Low estimate (at assumed rate of 40 per cent per annum)		2209
Estimated United States market for colour televisions	3548	4067

Source: Based upon Table 4.1 and *Electronics International*, January 1981.

4.2 Sectoral distribution

Most of the existing turnkey vendors had their origins in supplying CAD systems for the electronics industry. In this sector CAD was initially used by the large semiconductor firms to assist in the design of integrated circuits, but, increasingly, new CAD electronic systems (rather than enhancements of existing systems) are being used in the design of printed circuit boards. While this involves the penetration of new, smaller firms, there is some evidence to suggest that the market for electronics systems is growing less rapidly than for other applications (Kurlack, 1980).

Currently the major market [6] is for mechanical systems where the penetration of CAD technology is low and the aggregate potential is high. A large potential also exists in the architecture, civil and structural engineering (ACE) fields, as well as in new fields such as retaining layout, business graphics and animation.

Because of the different origins of the various vendors, each has historically developed a speciality in a particular area (see Table 4.3). But because of the directions of market growth those suppliers with a poor base in the manufacturing and ACE areas are scrambling to cover these sectors, often being forced to buy-in software and to integrate it into their own graphics and applications[7] software. Moreover many users demand an all-round capability of their CAD systems – for example they should not only be able to undertake mechanical stress analysis, but also be used for printed circuit board layout – and this, too, is forcing most vendors into developing a multisector applications program base.

Table 4.3 Estimate of sectoral distribution * of sales (per cent) of major vendors

	Electronics	Mechanical	ACE	Mapping	Other
Applicon (1980)	47.2	32.8	9.8	0	9.8
Auto-trol (1980)	10	36	50	22	18
IBM † (1980 and stock)	0	100	0	0	0
Calma (stock)	70	17.1		12.9	
Computervision (1981)	20	70		10	
Gerber (1980)	43	57	0	0	0
Intergraph (stock)	0	0	51	49	0
Unigraphics (1980)	0	100	0	0	0
Total	21.4	53.5		25.1	

* major end use.
† excludes internal use of electronic software.
Source: Interviews and Annual Reports

4.3 Geographical distribution

The market for CAD equipment began in the United States, with little external demand until around 1977, after which time the European and Japanese markets began to grow more rapidly than the United States market. Almost all vendors who were visited and who sold in Japan saw it as their fastest growing market; all reported an annual sales growth there of over 100 per cent per annum. As a consequence most of the vendors have become increasingly multinational in their sales and service and Computervision, as always the industry leader, is actively searching for a European manufacturing base. The geographical distribution of sales for various companies, where available, is shown in Table 4.4 below; but this masks the growing proportion of foreign sales – for example the proportion of United States sales for Auto-trol fell from 96.2 per cent to 77.1 per cent between 1975 and 1980, and from 84 per cent to 68 per cent for Computervision between 1971 and 1980.

Table 4.4 Geographical distribution of 1980 sales (per cent)

<table>
<tr><th></th><th>United States</th><th>Europe</th><th>Japan</th><th>Other</th></tr>
<tr><td>Applicon</td><td>70</td><td colspan="3">30</td></tr>
<tr><td>Auto-trol</td><td>77.1</td><td>17.3</td><td colspan="2">5.6</td></tr>
<tr><td>Computervision</td><td>68</td><td>22.4</td><td colspan="2">9.6</td></tr>
<tr><td>Gerber</td><td>60.7</td><td>23.4</td><td colspan="2">15.9</td></tr>
<tr><td>Intergraph</td><td>79.8</td><td>16.2</td><td colspan="2">4</td></tr>
<tr><td>Unigraphics</td><td>80</td><td>20</td><td>0</td><td>0</td></tr>
</table>

Source: Annual Reports and Interviews

Because of our specific interest in LDCs it is useful, at this stage, to list the sales of CAD equipment (by the United States turnkey vendors visited) to LDCs; this is done in Table 4.5, but the vendors are not identified due to the proprietary nature of this information.

There are a number of relevant observations which can be made about these sales to LDCs. First, in total they comprise of around forty-four systems, or around thirty-two if South Africa and Yugoslavia are excluded. This is in the context of a global installed base (of the turnkey systems investigated in this study) of over 8,000 systems.[8] Second, most of the vendors had lost track of these sales to LDCs and were unable to service them; in one extreme case hardware had been shipped to Zaire some years ago without any supporting software and the equipment was obviously useless. Third the cost of servicing LDC markets is unusually high. One firm which is making an effort to penetrate the Brazilian market due to the future expansion of the automobile sector, is having to specifically trains software and hardware engineers to support the sale of an initial system; in another case the purchasing TNC petroleum company has specifically trained supporting staff at its own expense. Fourth, most LDC applications are for mapping, which is generally

Table 4.5 Sales of CAD systems to LDCs by major United States turnkey suppliers

Vendor Code	Number of systems	Date of sale	Country	Use	Type of user
A	2	?	India	Electronics (pcb's)	post office
B	0	—	—	—	
C	2	?	S. Africa	Construction, mining	indigenous firms
	3	?	Yugoslavia	?	state
	1	?	Zaire	?	?
D	3	?	Brunei, Sarawak, Oman	Oilfield mapping	TNC
	1	?	Argentina	Military	state
	1			ACE	indigenous firm
	1		Honduras	Mapping	state
	1		Chile	Mapping	state
	1		Mexico	Mapping	state
	3		Venezuela	Oilfield mapping	state
E	1	?	Argentina	Mechanical	TNC
F	1	1981	Brazil	Mechanical (automobiles)	TNC
	2	?	S. Korea	?	?
	1	?	Hong Kong	Electronics	indigenous bureau
	1	?	Taiwan	Iron-works	indigenous firm
	2	?	Iran	Out of operation	?
	1	?	Mexico	?	?
	1	?	India	?	?
	3	?	Yugoslavia	?	?
	4	?	S. Africa	?	?
G	1	1981	India	Mechanical	indigenous firm
H	7*		Brazil	Mechanical (aircraft)	state

* Comprises 14 per cent of 1980 sales of around 50 systems.
Source: Interviews

used in petroleum fields or has strategic counter-insurgency imperatives underlying its use. (In one of these cases the purchase was aided by funds from the United States Agency for International Development). Many of the CAD suppliers reported a significant degree of *general* interest from LDCs (especially China, Hong Kong, Taiwan and Singapore) concerned with the purchase of state-of-the-art technology, but without any *specific* knowledge of the uses of CAD, or even of its nature and cost.

It is important, here, to re-emphasize that this data on LDC use of CAD refers only to the sales of the major United States and United Kingdom turnkey vendors which were visited (and who account for the overwhelming share of global turnkey systems sales). It is possible, therefore, that a number of isolated sales have been made to LDCs of small dedicated microcomputer-driven terminals or of pure software-based systems. A further possibility is the development of CAD systems within LDCs themselves. To the best of our knowledge only one set of LDC CAD equipment exists and that emanates from the Tata Institute of Fundamental Research in India. It is worth emphasizing the technical excellence of the Tata Institute which already, in the 1950s, had built its own digital computer and was, then, abreast of technological developments in the United States.[9] There are currently four imported turnkey systems in India, three of which are installed for predominantly electronic applications. Working with the basic graphics software of these turnkey systems, the Tata Institute has assembled four further systems, two of which have been sold to research institutes. One of these runs off a minicomputer (with two terminals) using bought-in software from the United Kingdom and the other two run off small mainframe computers largely using Tata applications programs. These programs are predominantly for the electronics sector and do not appear to have the sophistication of their turnkey counterparts. For example, the printed circuit board applications programs have no autorouting facilities (which is standard on the turnkey systems and the dedicated microcomputer driven systems in the DCs) and operates via a discrete interaction between the designer and the terminal for each placement. Their basic operating systems and graphics software also do not appear to be as sophisticated as the turnkey counterparts, since the Tata CAD systems only runs two terminals off a minicomputer, whereas most of the turnkey minicomputers are able to drive between four and eight terminals each. Nevertheless, despite these deficiencies, the Tata systems do work – the major obstacle to wider diffusion probably relates more to lack of demand from an unsophisticated industrial base than the inadequacy of their own software.

In conclusion, therefore, it appears as if LDCs have so far made very limited use of CAD technology. There are some signs that CAD vendors are beginning to see the potential in LDC markets (especially in mapping, but also in mechanical engineering) but at present almost all suppliers are heavily preoccupied with catching up with demand in DCs and it will be some time before attention will be given to exploiting LDC markets. LDC CAD capacility seems limited to India and even there its development is embryonic [10] and links with the industrial system are weak.

Notes

1. Which are particularly strong in relation to sales of software rather than turnkey, packaged systems.
2. Of which little is known.
3. With the exception of the United Kingdom, where *some* firms reported that their British sales *growth rate* looked likely to decline. In one United States survey (Kurlack, 1980) 35 per cent of responding user firms intended to expand their number of CAD systems by more than 100 per cent within one year, and another 8 per cent of respondents aimed to expand by 50–100 per cent in the same period. However, in the latter half of 1980 the increasing incidence of discounts being offered in the United States market suggests to some extent a softening market in the face of continued recession (see *Harvard Newsletter* on computer graphics, Vol. 3, No. 10, 18 May 1981). See also Kurlack (1981), who projects an increasing downturn in the growth rate for 1982.
4. According to one estimate the demand for such low-cost draughting systems in the United Kingdom alone is around 17,000. Selling at a low price of $35,000 each, this provides a United Kingdom market of around $600 million with a global market of over $10 billion (based upon the United Kingdom's proportion of global sales in semiconductors).
5. Kurlack (1981) forecasts a reduction in the 1978–1980 growth rate over the 1980–1982 period. For this reason we estimate the 1984 sales as being somewhat less than $4 billion.
6. Considered on an annual basis, since the early growth of CAD sales to the electronics sector makes it the major sector for installed systems.
7. For example, Auto-trol, with a strong past track-record in mapping and draughting bought in basic 3D mechanical software from MCS and then invested around 70 person years of software input into integrating and upgrading it.
8. One detailed industry survey estimated that the installed base of non-IBM systems was 4,265 at the end of 1980. IBM expected to ship 1,200 systems to external customers in 1981 alone which suggests that 8,000 is a conservative estimate of total sales.
9. Which built the first digital electronic computer in the 1950s. The Indian version however was non-electronic and worked with vacuum tubes.
10. The total stock of software input for the Tata Applications programs is less than 30 person years, which is very small by comparison with the turnkey vendors (see Table 5.4).

5 THE CAD INDUSTRY: MARKET STRUCTURE, MARKET PERFORMANCE AND NEW ENTRANTS

In the following three chapters we shall analyse CAD technology from the vantage of downstream users. But before we undertake this exercise it is first necessary to understand the organization of the CAD supplying industry. The issues that concern us here are the nature of market growth (i.e., size, and the sectoral and geographical distribution of sales), market structure, market performance, barriers to entry and the scope for new entrants in the industry. Through this analysis we shall not only gain a perspective on the possibility of LDC firms entering high technology sectors such as CAD, but also the question of whether suppliers are likely to market the technology actively in LDCs. Those readers interested primarily in the international location of industry might wish to skim this chapter which covers the relevent issues in some detail. The major conclusions, however, are relevant to subsequent discussion and will be drawn on in later chapters.

5.1 Market structure

Potentially the market for CAD systems is segmented into three major divisions, namely:

(a) mainframe-based systems which are partly used for graphics and partly for information processing and analytical tasks;
(b) minicomputer-based systems dedicated to graphics use, which can undertake a variety of different applications programs and with limited ability to perform analytical programs; and
(c) microcomputer-based systems with a basic draughting software and dedicated to a few applications programs with a very limited analytical capability.

Each of these systems could be made available by turnkey suppliers, selling complete systems of hardware and software, or by specialized software vendors which provide software alone. However the running so far over the past decade has been made almost entirely by the minicomputer-based turnkey systems, with the exception of IBM (whose presence[1] has hitherto largely been limited to very large companies each using a relatively large number of graphics terminals with the processing mainframe also being used on a batch basis for additional heavy analytic and data processing activities) and the emerging sales of software systems such as those of Cambridge Interactive Systems[2] and Compeda[3], both of the United Kingdom.

Currently seven United States firms dominate this turnkey market[4] – the shares of these individual firms are shown in Table 5.1 below for the period 1976–1980. The most significant factor which emerges from this tabulation is the growing dominance of the market by Computervision whose share grew from 28.1 per cent in 1976 to 33.2 per cent in 1980, at the expense of all of the other turnkey vendors except Intergraph. It is significant that the extraordinarily high growth rate of the industry means that despite a 78.4 per cent annual growth in sales between 1979 and 1980 a firm like Auto-trol was faced with a declining market share, or that Gerber, with a 31.4 per cent per annum growth rate over the same period[5] must be seen as struggling for survival! Indeed, according to Kurlak (1981), both of these firms were running at a loss by late 1981.

In order to understand the pricing policies and marketing strategies of particular firms it is essential to recognize the distinction between the marginal and average cost of CAD systems. All major turnkey vendors are concerned to extend the quality and range of their software – leading, as we shall see, to the employment of many expensive software writers. The software costs are fixed, that is they are largely incurred irrespective of the numbers of systems actually sold. Financing these software development costs and providing fixed and working capital to cover sales growth[6] implies an average system price (i.e., including a provision for software development and growth) which substantially exceeds the marginal cost (i.e., the hardware component) of each system sold.

Without exception the pricing policy of turnkey vendors is to load these costs on to hardware prices – that is, to 'overcharge' for the hardware in their systems (which, they argue, includes 'bundled' software) rather than to charge users for a realistic share of software development costs. Thus, as most users visited were at pains to point out, they were being forced to pay significantly higher prices for hardware than they could obtain by unpackaging the purchase of the system and buying in the hardware themselves. For example, one user could have purchased a disc drive from the original manufacturer for $8,000, whereas the list-price for the identical equipment supplied by the CAD vendor was $75,000. Two vendors visited freely acknowledged this and volunteered that they on-priced hardware at 'more than one-and-a-half' and 'more than two times'

Table 5.1 Market shares (per cent) and controlling shareholdings of major United States turnkey CAD suppliers

Market shares (per cent

	1976	1977	1978	1979	1980
Applicon	14.5	15.1	10.9	9	8.8
Auto-trol	10	11.4	13	10.6	8.9
Calma	14.3	15.5	16	13.6	13.8
Computervision	28.1	25.6	25.8	32.6	33.2
Gerber	3.6	1.3	2.6	3.2	4.1*
IBM	21.4	18.2	13.6	12.7	12.2
Intergraph	8.2	8.3	12	9.3	9.8
Unigraphics			3.3	2.4	2.3
Other					6.9

* Excluding estimated sales of PC800 dedicated electronic systems sold by a different division, Gerber's share falls to 2.3 per cent in 1980.

Shareholdings

	Control	Other significant
Applicon	Original founders until taken over by Schlumberger in 1982	General Electric owns 28 per cent; will divest due to purchase of Calma.
Auto-trol	Hillman Foundation (ca. 75 per cent)	Employees (25 per cent)
Calma	General Electric (100 per cent)	
Computervision	Founding president and vice president (22 per cent)	Institutions (38 per cent)
Gerber	Gerber Scientific (80 per cent)	Public (20 per cent)
IBM	Public 100 per cent	
Intergraph	Eight directors and employee fund † (40.6 per cent)	Public (13.2 per cent) Employees (46.2 per cent)
Unigraphics	McDonnel Douglas (100 per cent)	

† Voting rights of employees fund held by founder and his wife.

the hardware *list* prices, which in themselves are around 30 per cent higher than the original equipment manufacturers (OEMs) prices that they paid for the equipment.

This pricing policy of on-pricing hardware, rather than charging more realistic prices for software, is an industry-wide practice and is justified by vendors for the following reasons.

(a) Customers are loath to pay 'high' prices for a non-tangible asset such as software.
(b) If more realistic prices are charged for software it would provide the opportunity for small-scale software houses to supply competing applications programs.
(c) Multiple-systems purchasers would object paying each time for software which they have purchased with earlier systems.
(d) It inhibits users trying to copy software for sales to unrelated parties putting together unpackaged systems.

For obvious reasons, CAD vendors are anxious to prevent unpackaging, that is the unbundling of hardware and software components of the system by users (referred to by one vendor as a 'cottage industry') and two major stratagems are pursued to inhibit this. The first is the universal policy of refusing to maintain [7] the software or hardware of any system which contains 'unauthorized' unpackaged hardware; the second is to modify the hardware or coded software in particular sets of hardware (for example the interfacing nodules of disc drives) which prevents the mating-in of independently purchased hardware.

Given this difference between average and marginal costs (which is a common characteristic of software-intensive industries), and given the spectacular potential of the industry, the competitive pressures towards discounting sales are intense in particular segments of the market – many users visited reported obtaining such discounts. There are a variety of reasons why this discounting occurs.

First, given the difference between average and marginal costs, there is plenty of scope for discounting. Second, some suppliers observed that in particular markets (e.g., currently Australia and Japan) vendors with an existing presence had set aside a predatory pricing fund to inhibit entry by newcomers. A third factor leading to discounting is the policy of particular firms who are facing relatively stagnant markets and aim to increase market share, a policy which Gerber is now openly pursuing (see Drexel Burnham Lambert Inc., 1981). The fourth incentive to discounting occurs because companies which are large potential users are an important catch for vendors, since moving to an alternative CAD system in the future is often an expensive business.[8] And, finally, the recession in the US economy in 1981 increased the competitiveness of the CAD industry due to a fall-off in the growth of demand; this led to a tendency of some firms to discount sales (Kurlack, 1981).

While the primary realm of competition in the past has been the possession of specific applications programs that competitors did not have, the growing all-round strength of most CAD suppliers in recent years has led to an increasing tendency towards price competition. But as the nature of CAD users has changed over the past five years from specialized electronics firms, familiar with the problems of software maintenance, to relatively ignorant users in the mechanical and ACE sectors, so the nature of competition has begun to change from the realm of prices alone (in some cases 'commissions' of over $60,000 being offered to purchasing officers) to that of prices plus marketing and service. Most vendors are expanding their marketing facilities by well over 100 per cent annum, particularly in Europe. In other cases, critical marketing personnel are 'head-hunted'.[9] This growing emphasis on marketing and service, rather than on price, has been forced on vendors as they have begun to sell to less knowledgeable users. The first knowledgeable users are generally able to de-bug software themselves and make fewer elementary errors in use; they thus require minimal support. But later sales to less capable users require 'hand-holding' relationships. Consequently Computervision, currently the major vendor to the mechanical engineering sector in Europe, has been forced to establish a European Productivity Centre as well as a European Service Centre to cope with demands from relatively ignorant users.[10]

A further competitive strategem pursued by most vendors is to 'oversell' immature software packages to end users. This has three major functions. The first is to 'capture' a final user before it commits itself to a competitor, given that the costs of changeover from one CAD system to another are often[11] prohibitively expensive. The second is that it is difficult to simulate real operating conditions in the software 'laboratory'. For example Intergraph found it difficult to 'test' their data base management system (which is now a major competitive strength for that company) within the corporation due to their relatively underdeveloped data base; it had to go to a data-base intensive user before it could be effectively de-bugged. And a third advantage of marketing immature programs is that the user partly incurs the cost of developing and de-bugging the programs.[12]

5.2 Market performance

For a variety of reasons it is not easy – or always meaningful – to estimate the relative market performance of CAD supplying firms since many of them are still at the early stages of development and are incurring substantial development costs. Moreover most of them have only recently gone public (and therefore been forced to publish balance sheets) and therefore the relevant information is not readily available. But in Figure 5.1 we compare the return on equity of two CAD firms – Computervision and Intergraph – with the performance of DEC, the most successful of the mini-computer firms, and with the largest 500 United States firms (as published in *Fortune* magazine) over the same time period. From this comparison it appears as if the two CAD vendors appear to be relatively profitable and that, by implication, CAD firms in general are more profitable than those in most other sectors.

However, the return on equity is not the most suitable indicator to measure market performance since the CAD sector is still in its infancy and many firms are experiencing heavy development costs. Much more relevant from the investor's viewpoint is the appreciation in the value of shareholdings. Given the extraordinarily high rate of growth of the industry (see Figure 3.4) and the continued profitability of CAD firms (Figure 5.1) it is to be expected that the value of shares will have increased dramatically. But it is worth detailing some of these gains to obtain a true grasp of the returns to investors:

(a) *Computervision.* The original value of Computervision shares was 5 cents each, at which price the two founders and major current shareholders (including Martin Allen the President, and Paul Villers the former Vice President) obtained most of their stock. The values of these shares over the years are shown in Table 5.2. By January 1981 3,706 shareholders held 13,347,593 shares, of which Allen and Villers accounted for around 2,936,470 (i.e., 22 per cent). Given that they purchased most of these shares at the original par value of 5 cents and estimating that their average purchase price was around 50 cents per share, the combined appreciation of the value of their shareholdings has been in the order of $200 million in 11 years.
(b) *Intergraph.* The founder, J. Meadlock and his wife, together owning 10.7 per cent of total shares (and holding the voting rights over the employees-fund holding of 14.6 per cent) paid around $120,000 for their initial holdings. At the 1981 issue price of $18 per share (par value of 10 cents) their *appreciation* was worth $21.9 million, and at the current share price of $30, the appreciation was $36.5 million.

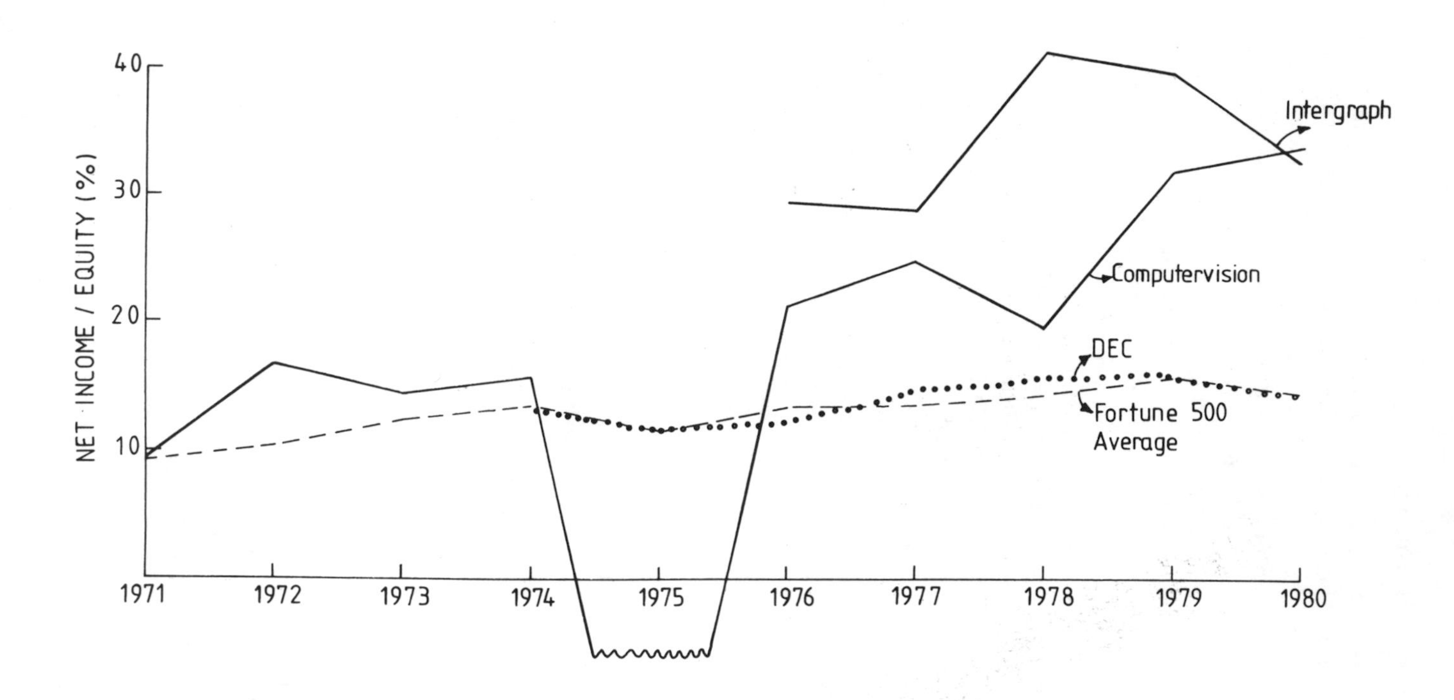

Figure 5.1 Return on equity (per cent)

Table 5.2 Appreciation in Computervision share prices

Date	Value of shares ($)	Notes
Original price (1969)	0.05	
4th quarter 1973 high	19.25	Recession in industry: see Figure 5.1
4th quarter 1974 high	5.13	
4th quarter 1975 high	5.50	
4th quarter 1976 high	5.75	
4th quarter 1977 high	11.40	
4th quarter 1978 high	27.63	
4th quarter 1979 high	56.38	
20 November 1980	50	
29 January 1981	70.12	

Source: Annual Reports and Merril Lynch reports

(c) *Auto-trol.* Auto-trol was purchased by the Hillman Trust in 1973 for $50,000 at 25 cents per share (plus loans at commercial interest rates of $3.1 million). This Trust was created in 1968 by Dora Hillman for the benefit of Howard Hillman and his children (Hambrecht and Quist, 1979). The firm was chosen by the Trust after an investigation into potentially 'explosive-growth' sectors for investment. At the 1979 share price high of $33.50, this holding was worth $67 million. In November 1980 the share price was $50, giving an appreciation of value of their holdings of around $100 million.

5.3 Barriers to entry

As we have seen, investment in the CAD sector is a relatively profitable activity, both in relation to potential income streams (although, because of the financial requirements for expansion, none of the CAD suppliers are known to have ever paid dividends) and in appreciation of capital stock. Consequently the incentive to new entrants is substantial. Yet given the small number of firms which dominate the industry, it is evident that there must be barriers to entry.

The primary barrier to entry in the CAD industry is technology since there are only insignificant economies of scale in production at the level of output of most of the relevant firms. Only IBM and Computervision make their own computers and even they buy in the peripherals from external suppliers. It is now widely accepted that technology is a 'free' or 'public' good, that is it is not used up in consumption and is available for multiple and indefinite use (Arrow, 1962). Therefore unless some mechanisms can be found to protect proprietary rights over technology, there will be no incentive, under a market system, for the generation of new technology. In the case of embodied technology – for example, machinery – this function is partly performed by the patent system which gives exclusive power for exploitation to the owner of the patent or its licensees. But, as has recently emerged in a series of well publicized court-cases, software cannot be patented,[13] it can only be copyrighted which is easily circumventable. Yet for the CAD industry to have maintained such dramatic changes in technology, there must by necessity have been a process of *effective* appropriation, whatever the *legal* rights may have been.

The primary factor protecting existing producers from new entrants has been the scale of software inputs necessary to offer a competitive package of applications programs. This would imply that the longer the industry has been in existence the more difficult it is for new firms to enter. While in the next section we will explore the potential avenues for new entrants, here we will confine ourselves to an assessment of the extent to which research and development (R&D) in software has erected barriers to entry. In Table 5.3 we detail the extent of R&D inputs of these firms for which information exists; in Table 5.4 we detail (without mentioning the names of firms to avoid disclosing proprietary information) the current numbers of software writers they employ and, where available, the accumulated input of software person-years in their system. It can be seen from Table 5.3 that compared to United States industry in general, the CAD industry invests a very large proportion of sales in R&D. This proportion is high even relative to information processing in general where the 1979 average was only 6.1 per cent of sales (*Business Week*) and the 1980 average was 2 per cent. From Table 5.4 it can be seen that most CAD suppliers currently employ over 100 software writers per annum, expanding these numbers at well over 30 per cent per annum. Some of the vendors were able to detail their stock of software – over 1,000 person years in some cases and over 7 million lines of code in others.

These absolute sunken R&D expenditures are in themselves not a sufficient deterrent to entry. After all, in 1980 prices, even Computervision had accumulated less than $80 million of R&D, and some of that was in the development of their own minicomputer; compare this with the $1.5 billion profits earned by General Electric in 1980 alone. The critical protective factors are that this software develop-

Table 5.3 R&D as a percentage of sales

	1969	1970	1971	1972	1973	1974	1975	1976	1977	1978	1979	1980
Applicon								11	8.4	12.8	8.9	11.2
Auto-trol *						22.8	13	14.6	12.6	15.7	14	12
Computervision	671	28	4.4	6.4	9.7	11.8	12.3	8.9	9.6	8.3	8.8	
Intergraph								6.5	5.7	7.6	10.1	11
Gerber												15.9
All US industry	2.2	2.2	2.1	2	2	1.9	1.9	1.9	1.9	1.9	1.9	2

Source: Interviews, annual reports, Soete (1977), Soete (1979) and *Business Week*, June editions.

* In 1980, Auto-trols ratio of R&D expenditure to sales (which at 12 per cent of sales, was the lowest ratio for the firm since 1974) was the fourth highest ratio of the 744 United States corporations surveyed by *Business Week*; its R&D expenditure per employee was the third highest of the sample.

ment occurs in a relatively specialized sector in the context of a general shortage of software writers. But more importantly, much of the necessary software development is sequential – as Paul Villers, one of the founders of Computervision described it, 'One can't make a baby in one month with nine women'.[14]

Table 5.4 1980 software staff and accumulated person years of software*

Firm code †	Numbers employed	Accumulated years of software in system
A	40	130 person years post bought-in package in 1974
B	150-200	1,000 person years
C	103	
D	345	
E	125	
F	120	
G	110	600 person years in 2D draughting package plus 400 person years in bought-in mechanical package
H	88	500 person years of software; 7 million lines of software.

Source: Interviews.
* Excludes personnel on hardware development, but includes those working on operating systems of minicomputers.
† Firm code is not in the same order as in previous tables.

Although these accumulated R&D expenditures are a significant form of appropriation and consequently an effective barrier to the entry of new firms, some CAD firms nevertheless take additional steps to avoid the disclosure of proprietary software information to competitors; this involves control over the software itself and over the movement of personnel:

(a) *Control over software*. Software programs can basically be divided into two segments – object code (usually referred to as machine code), which is made up of a string of 0's and 1's (i.e., binary code) and source code which is a higher level of programming language and which assembles the object code into less cumbersome bundled of instructions. None of the CAD firms objected to the release of object code to users (since it is an essential requirement for the functioning of their hardware) except for IBM which does not market its electronic CAD software. But significant differences were displayed in the attitude of firms towards the release of source codes. Four of the CAD supplying firms had no objection to its release, although one observed that the source code itself would be inoperable without a particular item of hardware over which it maintained tight and exclusive control; a second of these firms had merged with a parent which had developed its own software, and observed that even though it had unfettered access to this software with the full co-operation of its affiliate, it took years to unravel; a third firm only released source code if the user entered a 'Proprietary Software Agreement'. The remaining firms retained tight physical control over the source code, despite their general observations that they would not feel unduly threatened if their competitors obtained access to it. In one of these companies, which employs over 100 software programmers, only four people have access to the full store of source code; in another case a user reported that the CAD suppliers applications engineers came and physically removed the source code which they had 'inadvertently obtained'.

(b) *Control over manpower.* The flow of manpower, holding firm-specific rather than individual-specific knowledge is an important concern for all CAD firms, although of diminishing importance as the stock of software grows and individual specific knowledge becomes increasingly differentiated from firm specific knowledge.[15] There are a number of notable examples of the gains and losses flowing from such mobility. In the mid-1970s, as we have seen earlier, a team of West Coast software writers prepared the basic mechanical applications package for Computervision having earlier performed a similar task for Gerber. When the 1975 recession forced economies on Computervision, the head office in the Boston area tried to regain control over these programmers and wanted them to return to the East Coast. With few exceptions they refused and the recalcitrants formed their own company, Systems Science Software, later taken over by Calma. Computervision responded with a law-suit alleging:

(i) unfair competition;
(ii) breach of contract not to compete;
(iii) contractual interference; and
(iv) 'conspiracy'.

Calma countersued for $127.5 million punitive damages (under the Sherman Anti-trust Act), but both claims were eventually dismissed without damages. However the upshot was a damaging delay (until 1978) in the development of Calma's own mechanical applications package since Calma had to ensure that it would not lay itself open to further litigation. A second example is that of the recent move (discussed earlier) by the ex-head of Computervisions European marketing division to Intergraph. And a third example is that of Gerber, anxious to establish a strong package of mechanical applications software, which has recently 'poached' twelve to fifteen software writers from Unigraphics which has developed over the years a strong suite of relatively bug-free mechanical applications programmes. Reflecting the importance, albeit declining, of this transfer of manpower is a repeated tendency to increase the costs to individuals of leaving the company. Intergraph requires its newly appointed European marketing manager to resell his stock to the company if he leaves their employment (Dean Witter Reynolds Inc., 1981, p. 20); the same company specifically sees its employee stock option and bonus plans as a way of reducing employee turnover. In another case of a CAD bureau, control was kept over one key software writer by the granting of a concessionary mortgage for the purchase of a house. But these safeguards are not always wholly effective: one user pointed out that whole sections of the software manuals of two different suppliers were almost identical, reflecting the mobility of personnel between these firms.

(c) *Firmware*. It is an increasingly common occurrence in the electronics industry for software-intensive firms to wire in particular sets of software into hardware. In part this reduces the input of programming, which both saves costs and speeds up processing time. But more importantly it 'hides' software from potential competitors, and following a recent United States Supreme Court ruling, unlike software, firmware is patentable. For example Intergraph has a major competitive advantage over its competitors in being able to insert and withdraw information rapidly from its storage discs; the substance of this advantage is wired-in to a separate set of hardware called a 'scanner-processor' which Intergraph makes itself under carefully protected conditions. The move to increasing firmware is a major area of effort by all of the CAD vendors.

In conclusion therefore it is clear that whilst legal proprietary rights over CAD technology are of little effect, CAD suppliers are nevertheless able to maintain proprietary control over their software. Applicon, which has two United States patents for hardware, recognises the limited utility of these and of patents in general:

> The company does not consider that its success will depend upon its ability to obtain and defend patents, but rather on its ability to offer its customers high-performance products at competitive

prices (Blyth Eastman Paine Webber and Alex Brown and Sons, 1980, p. 19).

Auto-trol comes to an identical conclusion:

The company does not hold any patents or licenses covering its graphics systems. It may be possible for competitors of the company to copy aspects of its systems even though the company regards such aspects as proprietary. However the Company believes that, because of the rapid pace of technological change in the electronics industry, patent protection is of lesser significance than factors such as the knowledge and experience of the Company's management and personnel and their ability to develop and market its products. (Hambrecht and Quist, 1979, p. 20)

In the face of this effective appropriation of technology, of the sequential imperative in software development and the minimum scale in input required to develop a wide-ranging set of applications programs, the barriers to entry are high (but not insurmountable, as we shall argue in the following section). A number of examples illustrate this:

(a) *General Electric*. General Electric is one of the largest industrial corporations in the world with a 1980 turnover of $25 billion, and is considered to be the largest user of CAD systems in the world. Its various divisions currently use over 100 systems (in the order of Computervision and Applicon, with Calma a poor third, being used predominantly for electronic systems) and plan to purchase an additional twenty-five systems per annum. In the face of increasing technology-based competition in all of its markets (see *Business Week*, 22 December 1980 and 16 March 1981) the company has made a major decision to upgrade its own technology and in particular, through association between its new semiconductor division (General Electric also recently purchased Intersil, a semiconductor supplier), its machine-tool division and the CAD affiliate (now all grouped in its Industrial Controls Group), to develop the 'factory of the future'. An essential ingredient of this was a CAD/CAM capability – instead of starting afresh GE tried to increase its 28 per cent speculative holding in Applicon, which had a particularly strong capability in mechanical engineering CAD software. When this failed it switched its attention to Calma which was a less attractive proposition since its historic strength lay in electronics applications. General Electric paid around $170 million for Calma in early 1981 (compared to the price of $17 million paid for it by United Telecoms two years previously).

(b) *IBM*. In the mid-1970s IBM made a decision to expand its mechanical applications CAD capabilities for two major reasons. The first was the need to extend sales growth in the face of declining hardware prices; the second was its own need for CAD/CAM applications as it struggled to keep abreast of competition within its own sector. Despite having around 150 person years of software in their system, they preferred to buy-in Lockheed's software package since it had a track record of successful mainframe-based users. Lockheed are reputed to currently have around 120 people upgrading and maintaining this package.

(c) *Gerber*. As we have seen, Gerber was an early entrant to the CAD/CAM market, having bought in the basic mechanical software from Systems Science Software in 1973, and investing over 70 person years of software to upgrade it by 1979. However their effort languished for some years and it was only when a new president was recruited from United Technologies in 1978 that their CAD activities began to expand significantly. This was followed by the separation of Gerber Systems Technology (now the CAD division) from Gerber Scientific in 1979, and the sale of 20 per cent of its equity to the public in March 1981. Since 1979 Gerber has been struggling to increase its market share through, by its own admission, extensive price discounting and the widening of software applications programs. But the costs of this have been substantial (Hambrecht and Quist, 1979). Just prior to the sale of 20 per cent of its stock to the public in March 1981, GST had:

(i) Accumulated losses (by December 1980) of $1.453 million and expected these to increase until at least May 1981.

(ii) GS had written off $4 million of debt by GST in exchange for 320,000 shares.

(iii) $2.081 million of the expected stock issue of $7.1 million would go to repay additional debt to GS.

(iv) Additional long-term debt of $4.5 million.

(v) These losses were incurred despite the sale of a technology licence to a Japanese firm (see later) which netted $1 million in advance, plus an annual minimum royalty on sales plus a share of its licensee's pre-tax profits.

5.4 New entrants in the 1980s

In discussing the future composition of the CAD industry, it is first necessary to come to a view on what major directions systems technology will take, although this is in itself partly dependent upon the decisions of potential entrants. A decision by IBM, for example, to commit itself to the CAD industry will have a very significant impact upon the industry's future structure.

Over the past decade, as we have seen, the industry has been dominated by specialized turnkey firms. Some of these (e.g., Calma and Computervision) originally began as suppliers to the electronics industry; others have always been strong in mechanical engineering applications (e.g., Applicon, Gerber and Unigraphics) and yet others developed with an expertise in mapping and draughting (e.g., Intergraph and Auto-trol). But all of these suppliers now recognize that as the market widens, specialisms are dangerous – Intergraph, Calma and Auto-trol[16], for example, are moving as rapidly as possible towards a suite of mechanical applications programs to complement their existing strengths. And all suppliers are at the same time trying to strengthen their suits of applications programs for the ACE sector, which they see as the major growth sector in the second part of the coming decade, the first part of which is commonly recognized as being fuelled by the mechanical engineering market. So in respect of the development of all round applications programs, there is little difference between the strategies of the various turnkey suppliers.

Nevertheless one of the existing suppliers has chosen a particular strategy that in part differentiates it from its competitors. Computervision has in the past few years been making its own minicomputer, whereas before it dual-sourced with DEC and Data General processors. There are a variety of reasons why Computervision began to manufacture its own equipment, namely:

(a) The policy of dual-sourcing, which was designed to prevent undue reliance on a single supplier and to reduce the incentive to either of them to enter the market, created difficulties in servicing.
(b) The technology for these minicomputers was well known; moreover particular parts of their operating system software architecture was sub-optimal from the point of view of graphic technology.
(c) Computervisions's volume (500 systems sold in 1980) was too small to persuade the minicomputer manufacturers to wire-in ('firmware') particular software instructions.
(d) Most importantly, own manufacture increased profitability as it removed the profit margin of the minicomputer manufacturers who were buying in their components from sources equally available to Computervision.

This move to original manufacture of minicomputers was a shrewd one and provided rapid returns. But the CAD industry is currently at a major turning point with respect to the computers used and in this lies a potentially severe obstacle to Computervision. That is, the industry is moving from earlier 16 bit computers to 32 bit machines which are much more powerful, provide substantially quicker and more accurate processing at a lower unit cost and, perhaps more importantly, have the advantage of allowing for the use of data-processing batch programs as well. They are, in effect, small mainframe computers and consequently provide all their advantages. Currently only one CAD supplier – Auto-trol – offers a 32 bit machine; but all competitors aim to do so within the next 12–17 months.[17] It is thus critical for Computervision that their own 32 bit machine will be competitive – the problem for them lies in developing an efficient operating system and it is an open question whether with less than 100 software programmers working on the new operating system, they can provide an effective processor as rapidly as the market requires.

So, barring reservations about Computervision's hardware policies, it is evident that all of the existing turnkey vendors are moving in a similar direction, that is to provide a comprehensive set of applications packages to users. They are threatened however from both the higher (i.e., the mainframe based systems) and the lower (i.e., microprocessor-driven systems) end of the scale.

(a) *Mainframe-based systems.* Hitherto only IBM of the mainframe producers, has provided a computer-graphics- (as opposed to computer-) aided design capability. The reason for this has been that CAD is too specialized a use – the mainframe manufacturers tend to go for economies of scale in applications packages – and that, in the past at any rate, the market has been too small to bother about. The attitude of IBM illustrates this well. IBM see the problem as one of developing an information-processing capability for the engineering sector – this includes stock and wage control, production planning, numerical control, heavy-analytical programs, and, *inter alia*, graphics. Within this, graphics has particular requirements, namely:

(i) Data entry is spasmodic compared to the regular needs of stock and wage control.

(ii) The variations in processing requirements require excess processing capacity which computer managers, especially in engineering, find difficulty in acknowledging.

(iii) Automation of design and production has very substantial processing requirements.

Nevertheless IBM sees graphics as a necessary part of an information-processing system and as a way of increasing the processing requirements of users. Such heavy data requirements require mainframe processing capabilities, which is IBM's competitive strength. For IBM the competitors in this field are not so much Computervision and company, but CDC, Honeywell, Hitachi and Fujitsu.

As manufacturing industry (and the ACE fields) moves towards greater automation so the extent of the market and the imperative to mainframe producers of offering a comprehensive package of applications packages (which include design graphics) increases. All the mainframe suppliers are therefore being forced to offer CAD software. CDC as we have seen, has bought in Lockheeds software; CAD and Honeywell have already licensed Hanratty's AD 2000 software and other mainframe firms are evaluating it. Undoubtedly, therefore, the mainframe firms will develop a greater presence in the CAD field in the future. But since these firms thrive on economies of scale, and since their mainframes are suitable for many other uses as well as graphics, they will inevitably have greater appeal for the large-scale users who will, where necessary, be able to develop their own specific applications software or buy in this software from emergent software suppliers.

(b) *Microprocessor-based systems.* The emerging space for microprocessor-based systems has been made possible by two factors. The first, and obvious pre-condition, is the development of increasingly powerful microprocessors. The second is the maturity of specific applications programmes.

In order to understand the significance of this latter point it is necessary to return to our earlier observation that there is a distinction between the average and marginal cost of CAD systems. This distinction arises, as we have seen, from the imperative to all CAD suppliers to employ a large overhead of software writers. These software writers are in general divided into three groups. The first is committed to the development of operating systems – obviously this is a particularly critical task for Computervision which alone manufactures its own computers and employs over 100 programmers in this area, but all vendors continually need to update and improve their operating systems. The remaining software writers are divided between a small number who upgrade and maintain existing applications programs and the larger number who are involved in developing the new applications programs which are necessary to keep an all-round presence in the industry. From each supplier's point of view, therefore, the software development process takes the form as represented by Figure 5.2: a constant overhead investment in software,

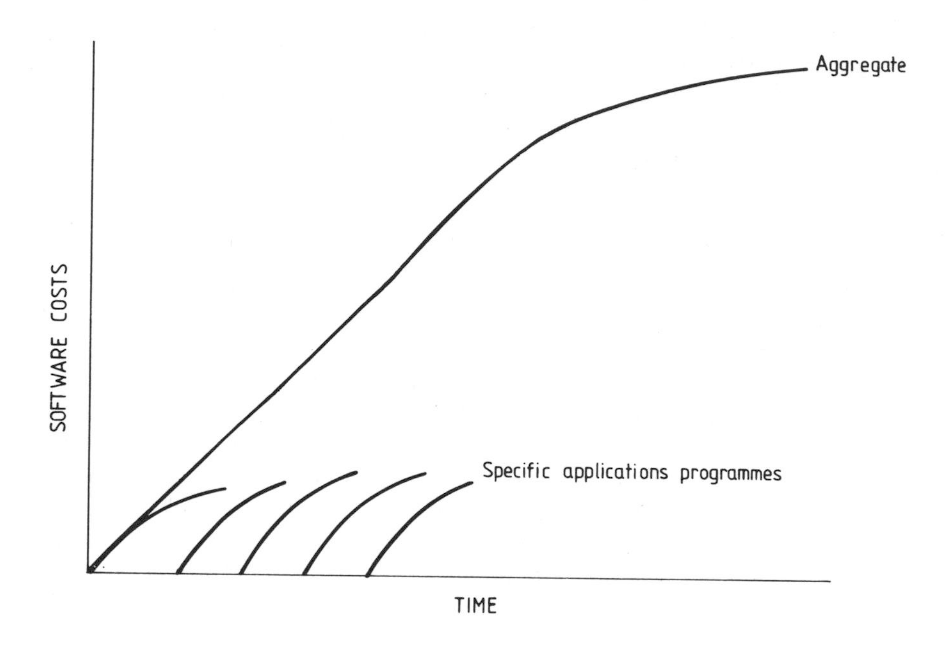

Figure 5.2 Software development – aggregate and application curves

made up of a family of matured packages and an increasing family of new packages. Evidence of the existence of this type of mature packages is provided in Figure 5.3 for three specific applications, sculptured surfaces and view-independent construction developed by one particular supplier, and orthographic piping developed by another.

Microprocessor-driven systems are aimed at this market of matured applications programs, of which basic graphics capabilities (i.e., the ability to draw lines, arcs and so on which are necessary for computer-aided draughting) is the most obvious. From the vendors point of view the software needs little attention and the systems can be sold at a price that is close to its marginal cost, in other words the cost of the hardware input.

There are a great number of such small firms springing up in North America (where according to one estimate there are at least sixty different suppliers) and Europe. Many of them offer basic graphics capabilities with perhaps one applications program, often an auto-routing electronics program. Many of these firms are started by former employees of established turnkey suppliers who recognized that these turnkey firms were overstretched and unable to satisfy

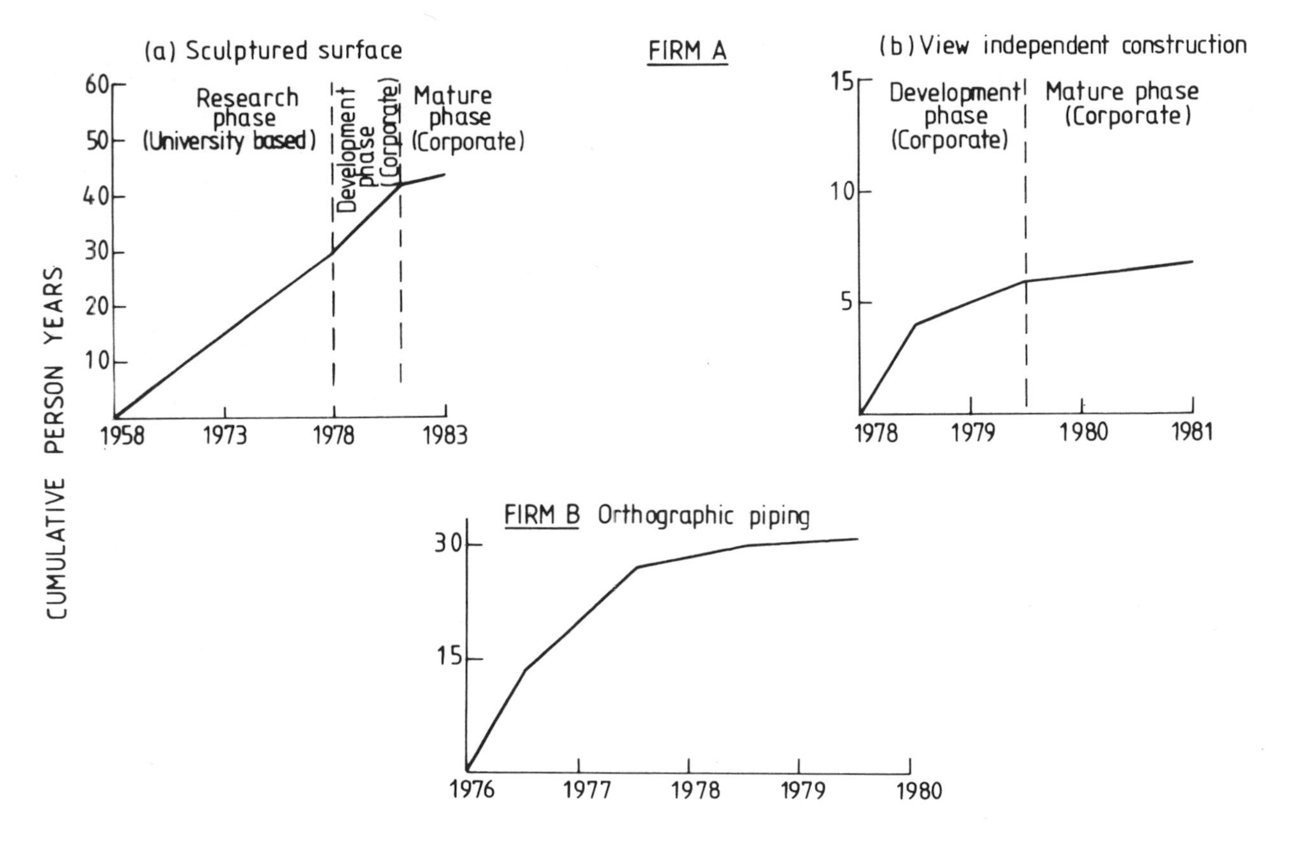

Figure 5.3 Examples of matured applications programs.
(Source: interviews with CAD supplying firms)

the needs of small scale users[18]. The weakness of these microprocessor-based systems is that they are non-upgradable (i.e., they can't be made more powerful) and that they come with a single, or a limited number of applications programs. They are likely to appeal to a small, highly specialized firm or as the equivalent of a typist's word-processor for a draughtsperson. It is in this latter application that the true growth area for these microprocessor based systems lies [19] – employers are more likely to provide a $40,000 terminal for basic draughting only than spend $100,000 for a terminal which is hardly ever used for the analytical work for which it is designed. In recent months, two-dimensional (2D) draughting systems linked to Apple microcomputers have been launched at a price of only $9,995.

An example of such an entry is a new system in the United States called Grafcon. [20] In 1977 a small architectural firm obtained a contract for wall pattern drawings from a major wood cabin building firm. The initial estimate of 2.5 hours per drawing turned out to be wildly optimistic – each drawing took around 10 hours in reality. So a microcomputer was purchased and specific graphics software was written, leading to a CAD turnaround time of 45 minutes per drawing within two months; the system also produced an error-free bill of materials as a bonus.[21] Realizing that it had a viable low-cost draughting tool on its hands an agreement was struck with A.M. Brunning, an old-established graphics equipment supplier, which was anxious to capitalize on a new technology which looked likely to rapidly eat away the market for its established products. Brunning negotiated exclusive rights outside of Southern America with an option to purchase it after three years. In 1980 alone thirty-seven systems were sold (worth around $1.2 million).

The existing minicomputer-based turnkey suppliers therefore face the potential of being squeezed at both ends of the market – from the high end by mainframes and at the low end by the microprocessor-based systems. 'Squeezed' however in the sense of market share rather than in terms of *aggregate scale of output*, since it is likely that the market for CAD equipment will grow very rapidly in the future. Only one CAD supplier – and none of the users visited in the course of this survey – felt that the aggregate software curve (see Figure 4.1) had matured; all believed that there would continue to be a role for new applications programs that are currently only being developed by the minicomputer-based suppliers. *Relatively* speaking, however (i.e., in terms of market share), the role to be played by the minicomputer-based suppliers will depend upon their ability to provide systems which can be used interactively with existing and potential mainframe suppliers, and/or to compete with the small-scale mass-draughting market. Few of the vendors visited seem to appreciate this potential problem.

The third major threat to the continued dominance of the industry by the existing minicomputer-based suppliers arises from the possibility of new firms entering the market with a comprehensive set of applications programs. Because of the effective barriers to entry discussed in the previous section, such an entry almost certainly would be based either upon the takeover of an existing supplier (as occurred in the case of Calma) or through the purchase of a licence on an existing software package. Of the existing turnkey vendors the most vulnerable is probably Gerber which continues to struggle for market share; but Computervision with its founders holding only 22 per cent and institutions holding around 38 per cent of its equity, must also be considered vulnerable to takeover. Potential entrants may come from as far afield as the petroleum industry (for example as with Exxon's plunge into word-processing and microcomputers), from a large United States industrial firm (as in the case of General Electric; or United Telecom, which, following its sale of Calma to General Electrics has just taken over Megatek, a large supplier of peripheral hardware to the graphics industry) or even from a European-based firm (such as Schlumberger [22] or Siemens) or Japanese-based firm anxious to enter this high-technology industry.

The alternative to taking over an existing CAD supplier is to licence basic CAD software and to develop it further. As we have seen many of the existing turnkey CAD suppliers pursued this strategy in earlier years. The most readily available set of such software is Hanratty's AD 2000 package which has already been purchased by Auto-trol and licensed by CDC and Honeywell. But such an entry strategy is not without its problems. The first is that however sophisticated packages such as AD 2000 are, they have not yet matured. One large-scale user undertook a full-length investigation of it in 1980 and concluded that if it were to 'procure AD 2000, a staff of thirty to fifty programmers must be dedicated to bug-chasing and coding enhancements'. Such a task was clearly beyond Tektronix, a manufacturer of CAD graphics screens, which attempted to enter the CAD industry by licensing AD 2000 but subsequently cut its losses and withdrew. It was not however beyond Auto-trol which has used AD 2000 to establish a suite of mechanical engineering applications programs. Interestingly Auto-trol originally took out a licence for AD 2000 and later converted this to an outright purchase (for $1 million) for two reasons. Firstly it was felt that royalty-based fees would prove more expensive in the long run than buying it outright. And secondly, Auto-trol had already put in around 70 person years of software into upgrading and de-bugging AD 2000 and was putting in more effort continually; this was considered proprietary information and Auto-trol were concerned that Hanratty's firm would benefit from its developments.

AD 2000 is only one of a number of potential software packages that might serve as an entry point to a newcomer to the industry. Another package, Pipework Design Management System (PDMS) was developed by a Dutch based firm (AKZO Engineering), the United Kingdom government-financed CAD Centre and an independent private British firm, Pipework. It is currently unique in being the only full 3D piping package [23] available and has an enormous potential role in the process plant fields. There are also a variety of solid modelling, 3D systems being generated (such as the French state-financed EUCLID, or the ROMULUS package offered by Cambridge Interactive Systems) which have an important future role to play in volumetric design and automated assembly.

And, finally, the existing CAD suppliers face the ever-present threat of Japanese entrants. Japan, as seen earlier is a major market, and is also the fastest growing single market for CAD systems sold by the United States turnkey suppliers. But there also appear to be indigenous Japanese systems – one United States CAD supplier remarked that Matsushita alone used seventy of its own terminals and software to design nuclear power plants. Perhaps more pointedly Gerber has entered into an agreement with a Japanese firm, Yokogawa Electric Works, whereby it obtained $1 million in advance, royalties on net sales (with a minimum of $1 million by 1985) plus a percentage of pre-tax profits from the sales of these systems plus free access to all Yokogawa-based developments [24]. In return Yokogawa has an exclusive licence to manufacture and market the systems in Japan, Korea and Singapore and a non-exclusive licence to sell in certain other non-specified countries. This agreement continues until 1988. And a third Japanese firm, Dai Nippon Screen, is beginning to diversify out of the printing equipment sector to electronics applications programs for CAD.

So far Japanese firms have not played a major role internationally in software-intensive industries, but it would be nearsighted to ignore their potential – certainly few of the existing CAD suppliers are unconcerned for the future.

Notes

1 However, IBM only markets its mechanical software. Its electronic software is not marketed outside of affiliates since IBM is concerned that this would divulge proprietary information regarding component, integrated circuit and computer architecture.

2 CIS markets, with increasing success, a 3D modelling capability and a basic draughting system. Annual sales are less than $10 million.

3 Compeda, a United Kingdom government-sponsored firm, markets a very successful 3D system (PDMS) to check interference between pipes in process plant design. Its annual turnover is less than $5 million.

4 We refer, here, to the market for CAD systems in the engineering (electronic, mechanical, civil and structural) architecture, retailing and publishing sectors. Excluded are the fields of business graphics (which according to one source – the *Harvard Newsletter* on computer graphics, Vol. 3, No. 9, May 1981 – was worth $200 million in 1981) and animation.

5 Excluding estimated sales of $10 million for the dedicated electronics system produced by an affiliate.

6 For example with revenues of only $131.6 million in 1979, Computervision in 1980 spent $31 million on capital investment, $22.1 million on research and development (R&D) and increased its working capital requirement over the previous financial year from $43 million to $72 million.

7 Which is generally covered by annual maintenance contracts averaging around 10 per cent of list prices.

8 Some CAD vendors admitted that many of these users misread the discounts which they obtained as resulting from the users' buying power: from the suppliers viewpoint it was a worthwhile concession to snare a captive customer in the future.

9 For example Intergraph have just recruited the former head of Computervision's European sales to spearhead their own expansion in Europe: the salary offered to this individual is not disclosed but he was given the opportunity to buy 45,000 shares at $4.17 each (Dean Witter Reynolds Inc., 1981, p. 22) just before Intergraph went public. At the market prices of these shares just after launching (i.e., $30), this marketing vice president was sitting on top of a gain in share value of $1,162,350!

10 Much of the criticism of Computervision's problems with software – especially in Europe – must be attributed to the fact that they have pioneered sales to relatively incapable users – that is, it is the users who are weak, rather than the packaged software they have purchased.

11 But not always. A number of the larger United States corporations visited made use of a variety of different CAD systems, sometimes even within the same office.

12 Once again this had two sides to it. One United States user had developed a simple applications program. Instead of selling this marketable software, it preferred to give it free of charge to the CAD supplier. The reason for this was that the user did not want to incur the cost of continually upgrading this software to make it compatible with the rest of the system, but it knew that once the supplier sold this software to other users it would be forced to maintain its compatability to the rest of the systems software.

13 However, a very recent ruling (May, 1981) of the United States Supreme Court had determined that 'firmware' (a form of software written into particular pieces of hardware – see later), can be patented. This is almost certainly likely to have a profound impact upon the direction of technological change in the whole of the electronics sector.

14 A similar example can be drawn from the automobile industry. When Agnelli, the head of Fiat, was asked whether selling the Fiat 124 design to the Russians would not provide fatal competition to his firm, he answered 'If we are still producing the Fiat 124 in five years time, we will go bankrupt anyhow'!

15 A major method through which CAD suppliers appropriate the individual knowledge of programmers is by insisting on the use of structured programming techniques.

These involve the use of standard procedures for software writing which are easily intelligible and assimilable by other software writers.

16 Through the purchase of basic capability from MCS called AD 2000.

17 In his survey of forty five United States CAD users, Kurlack (1980) found that forty would have preferred 32 bit machines.

18 For example Avera Inc. was set up by two ex-Applicon employees (one had worked at Applicon for seven years) and one ex-Intel employee. Set up with $150,000 of capital the single-terminal systems are sold for $39,250. The ex-Applicon founder described its origins as 'While at Applicon, request-upon-request would come in for design automation equipment priced at $50,000 or less' (*Harvard Newsletter* Vol. 3, No. 4, February 1981).

19 As we have seen, according to one estimate the market for such systems in the United Kingdom alone is around 17,000 terminals worth $600 million. This suggests a global market of around $14 billion for basic draughting aids.

20 See *Anderson Report*, Vol. 3, No. 81, April 1981. Additional information obtained through interviews.

21 In one case in the pre-CAD system a faulty bill of materials had led to the short delivery of one log to a distant site: rectifying this error cost of $15,000.

22 Schlumberger purchased Manufacturing Data Systems Inc. (for $189 million in 1980) which markets CAM software: the logic of linking this to a CAD capability (beyond the 2D draughting package currently offered by MDSI) was shown in the takeover in late 1981 of Applicon.

23 Amongst other things this package automatically calculates whether any two sets of pipe in a complex refinery design interact with each other or any other piece of equipment. This provides substantial savings in construction, as well as in design.

24 As a sign of the future Yokogawa rapidly developed auto-routing electronic software which Gerber now offers as part of its system.

6 THE BENEFITS OF USING CAD

The previous chapter was devoted to a detailed discussion of the evolving nature of the CAD supplying industry. This sets an important backdrop for subsequent discussion (in Part III) on the likely impact on LDCs of CAD technology and other types of electronics-based technologies. In this chapter (and in the two which follow) we address the impact that CAD has on the competitiveness of innovating firms. In focussing on the 'benefits' of its use, we refer to the gains accruing to the innovating enterprise and not that accruing to the people whose jobs are affected (positively or negatively), or displaced by the new technology. While these latter impacts are intrinsically important (and are in fact discussed at various points in the text – see also Arnold and Senker, 1982) they do not directly affect the impact of electronics related technologies on the international division of labour in industry. Indirectly they may nevertheless be important since, if the negative impacts on labour affect the nature of the technologies' diffusion, then this will have important implications for LDCs. Where relevant, therefore, these indirect impacts are also discussed.

It is, of course, almost impossible to compute the benefits of using CAD equipment. There are a number of reasons why this is so. As a general observation, the productivity of most technologies can seldom be measured as thing-in-itself since it will inevitably reflect the institutional and economic structure in which it is innovated; few plants ever perform at the anticipated rate of output. Such unpredictability of performance is particularly prevalent when the gains are based upon systemic reorganization, as is the case with CAD.

But there are other reasons, specific to CAD technology, which make it difficult to assess the economic benefits of its utilization. First, as we have already pointed out in the earlier discussion on the relevance of this study, CAD is a necessary and important element in the jigsaw which is leading to the emergence of the automated factory. In this change in industrial organization the real productivity gains lie in systems, rather than within particular sub-processes such as machine setting or draughting. Consequently many of the benefits of CAD will be felt downstream in production and information control and are difficult to quantify. Second, many of the benefits of CAD are realised in the quality of the product and the speed with which new or modified products are launched – these product improvements, often critical to the survival of particular firms, are difficult to measure and vary in importance between sectors and firms. And third (as in the case with software-writing in general – see Kraft, 1977) despite the strenuous efforts of management, much of design is an art form, highly dependent upon the particular individuals involved and consequently open to variable levels of performance.

It is not possible, therefore, to provide a blanket assessment of the benefits of using CAD since it so largely depends upon the particular circumstances in which it is used, and the organization within which it is innovated. The most we can do in this study is to illustrate the types of benefits that individual users have realized and attempt to classify these in some preliminary way. In the next chapter we attempt to assess which sectors CAD technology will most affect, and how this will affect LDC exporters of manufactured goods.

On the basis of interviews conducted with users in the United Kingdom and the United States[1] it is believed that the benefits arising from the use of CAD can be classified into three major areas, namely CAD as a draughting tool (i.e., the choice of technique), CAD as a design tool (i.e., both the choice of technique and its influence on products) and the downstream benefits that it allows (i.e., CAD as a key to systems gains). We will discuss each of these in turn as well as a series of other benefits relevant to the impact of electronics on LDCs before we turn to an overall assessment.

6.1 CAD as a draughting tool

There are three factors at work in determining the productivity that CAD systems realize. Firstly there is the speed with which the technology actually operates: this is not just the speed (and precision) at which the lines are physically drawn on the screen, but includes the routine background calculations which lie behind the drawing. For example, changing the specifications of a biscuit machine from one size to another, as occurred in the case of one user, requires the scaling of all components of the design: manually, this involves endless hours of routine calculation whereas on the CAD system used, the scaling is done automatically and almost instantaneously. A second and related factor is that it depends upon the type of drawing involved. Most users felt that on primary design drawings, the productivity of the CAD system was seldom higher than manual systems, but on modifications the productivity gains were higher, frequently over 20:1 and in one case computed at 100:1. And third, many of the gains from CAD depend upon the system with which the design office is organized. This is particularly important when the product is of a modular nature and particular parts can be stored in the memory of the CAD system – instead of manually redrawing a particular hopper or conveyor-belt in a plant design, the CAD operator can instantaneously recall it (or a variety of alternatives) from memory and append it to the layout under design.

Consequently there is an enormous variation in the productivity of CAD systems. Nevertheless a number of users visited had actually computed the overall average productivity of their CAD systems and the results are shown in Table 6.1 below. In most cases these productivity ratios were realized despite the fact that CAD was being used primarily as a design tool, rather than as an automated draughting aid.[2] On average, most of these users were realizing productivity gains of over 3:1 over manual systems. In some cases this has led to the displacement of existing staff but in other cases users reported that they preferred to realize a proportion of these benefits by producing more drawings (or more complete drawings) than they used to produce before CAD was introduced. One firm had displaced 25 per cent of its draughtspersons and remarked that the trade unions were 'suckers'; displacement in the future would be even higher once designs were stored in memory. (Subsequent to our visit a further 15 per cent of design and draughting staff were made redundant.) Another large-scale user estimated that it would displace around two thirds of its draughtspersons before 1985. A number of other firms found that CAD systems enabled them to make much less use of contract draughtspersons (who cost around double permanent staff). Indeed the displacement of contract work appears to have been the major area in which the job displacing characteristics of the technology have been felt.

Similar results were obtained by a survey of thirteen users in the United States and fourteen European users, which was conducted in 1978 (see Department of Industry, 1978). While no overall figures were made available by the users visited in that study, the productivity ratio reported varied between 2:1 and 27:1.

These productivity ratios depend critically upon a change in the organization of the design process and particularly in a change in the nature of work. For management, designers and draughtspersons have always been problematical staff to employ since their performance has been so individual-specific – none of the users had ever

Table 6.1 Productivity of CAD systems

Sector of activity	Location	Primary use of CAD	Average productivity ratio	Range of PR between different types of drawings
Integrated circuits	US	Design	2:1 after 6 months	NI
Automobile components	UK	Design	3:1 after 12 months	NI
Plant design	UK	Draughting	3:1	1:1 - 20:1
Process plant	UK	Design	NI	1:1 - 50:1
Electric motors	UK	Draughting	6.6:1	NI
Printing machinery	UK	Design/draughting	>2:1	NI
Architecture	UK	Design	3.5:1	NI
Automobiles	UK	Design	3:1	NI
Computers – pcb's	UK	Design/draughting	>5:1	NI
Process plant	UK	Design/draughting	4:1	NI
Petroleum exploration	UK	Design	2:1	NI
Automobiles	UK	Design/draughting	2.78:1 after 6 months	NI
Aircraft	US	Design	2.5:1 in 1979	NI
			3.32:1 in 1980	NI
Instruments – pcb's	UK	Design/draughting	>3:1	NI
Public utility	US	Draughting	>3:1	

NI = No information.
Source: Interviews with users.

been able to institute any form of 'Taylorism' of work-performance measurement. CAD in itself offers little to management to circumvent this, but many users – including draughtspersons themselves – reported that it significantly affected the pace at which they worked. This occurred for three reasons. Firstly drawing offices are social environments – many managers lamented that, on average, draughtspersons only spent around 30 per cent of their time actually drawing, with much of the rest spent in discussion. By contrast the CAD screen is a solitary work environment and operators tend to spend more of their time actually working than they used to at their drawing boards. Secondly, most of the available CAD systems offer reasonably rapid response and – in the interests of 'user friendliness – tend to prompt the operator. The consequence is that the design/drawing process becomes machine-paced rather than worker-paced, and is consequently speeded up. And finally, when each draughtsperson had $2,000 of equipment to work with, fixed costs were small and there was little financial incentive to work multiple shifts. But when CAD operators are backed by over $100,000 of equipment each, it becomes imperative for design office management to ensure full utilization, and most CAD systems – especially in the United States – are worked on a multiple shift basis.[3] Even in the UK a number of design-offices had moved to extended single-shift flexitime (around 10–12 hours per day) compared to single-shift working in the days of manual design.

As far as the 'quality of work' is concerned it is not easy to come to a clear conclusion. Most CAD operators interviewed (generally, but not exclusively, men) felt that the quality of the work environment had improved.[4] But this could have been a self-selecting group since the dissatisfied may well have left. One major United States user had lost twelve operators in the past year, at least half of whom had left because they disliked the move to multiple shifts. But it depends very much on the particular individuals involved – this was very clearly the case in one two-terminal design office visited. One of the operators enjoyed using the system as it speeded up the boring drawing involved in design. The other enjoyed drawing and saw it as a craft. He observed that whereas in the manual office the quality of his lettering stood out as an individual attribute, in the CAD era everyone's drawings looked the same. Moreover he lamented the loss of a more social work environment and illustrated this with a manual drawing that was annotated with various notations whenever he had been interrupted by colleagues. With CAD, he complained, he was never interrupted.

Given the sorts of productivity ratios observed, the costs of using

CAD equipment and the salaries of draughtspersons, it is possible to compute some of the economic benefits arising from the use of CAD. One United States user in the aerospace/defence sector had undertaken a detailed internal study of matching CAD and manual systems involving an equivalent work-load of 43 person years of manual draughting in 1980. Using the productivity ratios generated in this study (2.5:1 in 1979 and 3.32:1 in 1980) and the costs of using turnkey CAD systems which were primarily purchased for design rather than draughting (and were hence more expensive than low-cost draughting aids), the economics of choice are illustrated in Figures 6.1 (a) and (b) below. Figure 6.1 (a) shows how long it takes to pay back the costs of CAD equipment [5] – with the 1980 productivity ratio of 3.32:1 and working on a double shift basis, it took less than 10 months. From Figure 6.1 (b) it can be seen that given the 1980 productivity ratio and the cost of the CAD equipment, it would pay to use CAD equipment when gross salary costs (i.e., including overheads) exceed $10,000 per annum.

Yet these calculations – derived from actual observation – arise from the use of expensive CAD Design equipment, rather than the emerging low-cost basic draughting tools discussed earlier in this report. On the basis of purchasing such equipment for $35,000,[6] assuming extra costs of $5,000 as a contribution to the purchase of a system plotter (which can be used for a large number of draughting aids) and $10,000 for space and services (additional to that required by a manual draughtsperson),[7] and assuming a three-year payback period and a loan at 15 per cent per annum, it is possible to simulate the benefits to use. This is done in Figure 6.2, from which it appears that at a productivity ratio of 3:1 it would pay to use such low-cost draughting aids on a single-shift basis when gross salary costs (i.e., including overheads) were more than $11,000 per annum and, on a double-shift basis, when gross salary costs exceed $5,500.

These financial benefits, which we have calculated, only cover savings attributed to the productivity of designers and draughtspeople. However there are a range of other financial benefits that arise from better quality drawings, and drawings that have fewer errors. One of the users designing process plant remarked that certain drawings are characteristically revised 8 to 12 times – in manual offices this implies a messy and error-prone product; with CAD these problems are largely removed. Almost all of the users visited felt that the benefit of more accurate drawings was one of the major advantages of its use.

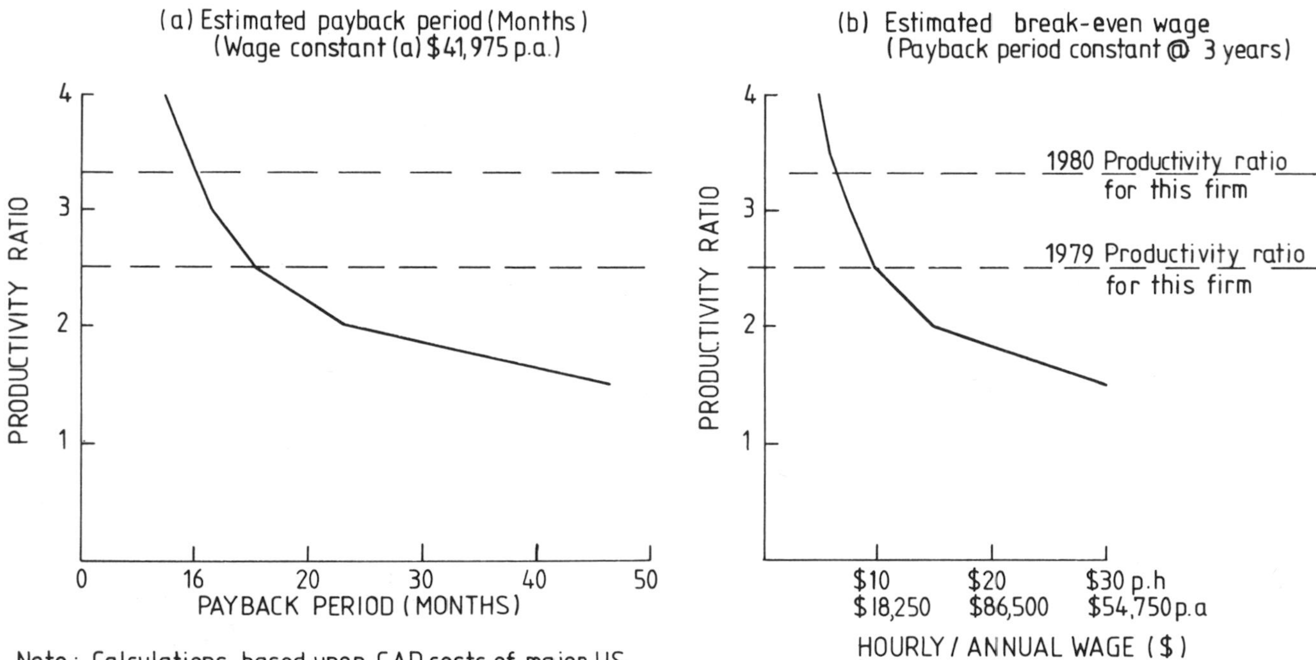

Note: Calculations based upon CAD costs of major US aerospace firm.
- terminal cost of $126,757p.a
- operator cost of $41,975 p.a
- 4 terminal system, working 2 shifts, giving 11,458 terminal hours p.a
- 1980 workload of 78,500 person-hours (=43 person years)

Figure 6.1 CAD design as a draughting tool – actual observation by large United States user

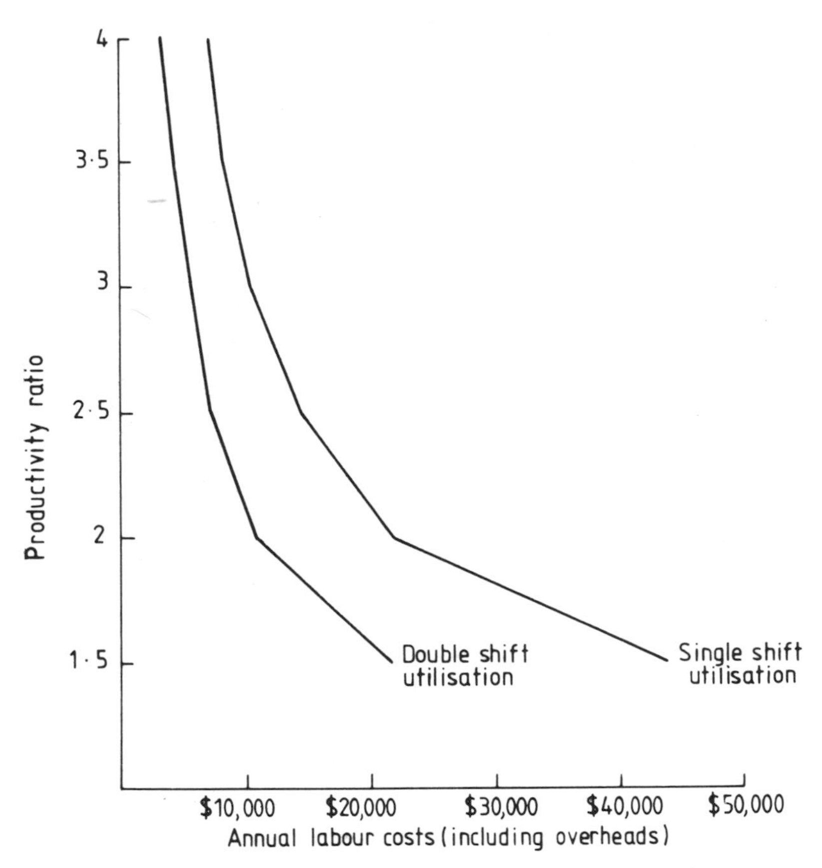

Figure 6.2 Hypothetical break-even wage for low-cost draughting system

6.2 CAD as a design tool

There are a realm of benefits that arise from the use of CAD that have an impact on the product that is produced. Here we can distinguish between three sub-sets of advantage:

(a) *CAD as an essential tool.* there are a variety of products that could not be produced without CAD such as VLSI integrated circuits, nuclear power plants, military radar and aeroplanes. While in many of these cases (with the possible exception of some electronics applications, where *visual* interaction is essential) designs could be produced with non-graphics-based computer-aided design, the absence of an on-line, interactive relationship (i.e., via graphics terminals) between designer and product would almost certainly make the final designs sub-optimal and uncompetitive in the market place.

(b) *CAD as an optimizing tool.* In an increasing number of sectors the differentiation and optimization of products is essential as competitive pressures hot up during the super-competitive down-swing of the long-wave cycle (see Chapter 2). CAD has an essential role to play in this, and almost all of the users visited felt that the *primary* function of their CAD systems was that it allowed them to modify and optimize their final products more effectively. This is particularly true in the 'capital goods' sector, loosely defined to cover large projects where the final product tends to be individually tailored to the customers needs and where design costs are a large proportion of total costs. An example drawn from the aircraft sector illustrates this well. In civil aircraft:

> R&D costs are shown to account for about half of the total launch costs. Design costs generally constitute around 25% of the R&D costs. Hence, 'design' accounts for about 12½% of the launch costs, and this is about 1% *only* of the selling price.
>
> However, during the spending of this 1% the cost effectiveness of the product is essentially defined. Design exerts a controlling influence on all elements of selling price and subsequent in-service costs, as well as having the responsibility for all aspects of performance achievement. (Jacobs, 1980, p. 178)

Figure 6.3 (drawn from British Aerospace studies of cost control) provides further evidence of these benefits. In it, the life cycle of an aircraft is divided into five phases, namely conception, validation, development, production and operation. The first two thirds of this life cycle is accounted for by design – *in this design cycle, while only 5 per cent of the total life cycle costs of the product are incurred, fully 85 per cent of the total lifetime operating costs of the aircraft are determined.* In this context, optimizing the final product is of critical importance, far more so than any savings in labour costs that might arise from the automation of design and draughting with CAD equipment.

Civil aircraft may appear to be an unusually clear case of the advantages derived from the optimization of product. But it is not unique. For example, a major sphere of competition in the automobile sector is associated with the need to make cars more fuel efficient. This involves an optimization of engine design, weight minimization and a reduction in the drag coefficient (Senker, 1980). All of these problems are currently being tackled with the assistance

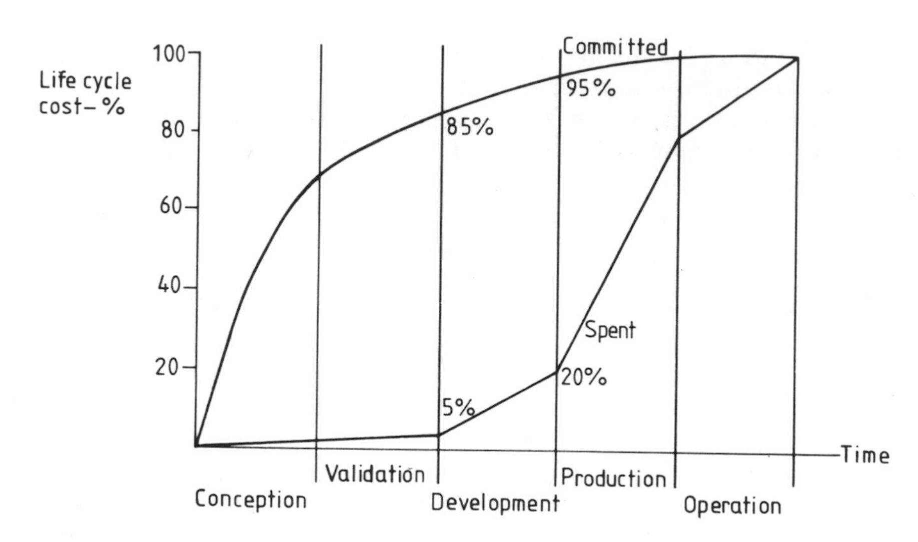

Figure 6.3 Life cycle costs – money committed vs. money spent (civil aircraft) (Source: Jacobs, 1980)

of CAD. Another rapidly emerging example concerns the layout of retailing premises and fast-food restaurants – one of the major chains of fast-food vendors explores alternatives and optimizes space utilization through CAD. And a final example points to the very substantial benefits that can arise from the use of CAD and that do not arise from productivity improvements in draughting. Increasingly, the boundaries between oil fields are defined by computered systems, and this is for a very important reason. In a typical North Sea oil-field, misplacing a boundary by 1 m can involve a loss of $60 million in oil – the oil companies consequently insist upon the use of computers in defining their oil field boundaries.

(c) *Product lead time*. While the cost of final products and their performance are critical areas of market competition, the timeliness with which products are launched is very often of even more crucial significance to particular firms. CAD plays a very important role in this regard through the reduction in lead times which it permits. This benefit is really the other side of the same coin and enhances the productivity (per labour hour) of design offices. While numerous examples could be given to illustrate the benefit derived from the shorter lead times that CAD permits (which almost all users felt to be one of the primary benefits of using CAD), two are particularly pointed. The first concerns a British firm that had been through various stages of asset-stripping ownership. Close to bankruptcy, one of their design engineers came up with a radical re-design of its major product, which had to be launched within one year or the firm would go out of business. On the basis of archiveal records it would have required 12 person years of draughting, but not only was there a shortage of draughtspersons in the area, but much of the design was sequential so the actual process would have taken at least 18 months – beyond the one year survival limit given to the first by its bankers. A two-terminal CAD system was procured at short notice and three designers produced 8,000 drawings in the first year, compared to the archiveal records of 400 drawings per person per year in the old manual office. In the second example of a British machinery manufacturer making effective use of CAD for modular plant designs, a designer was telephoned at 6.00 a.m. on a Saturday morning by a salesman about to go to Mexico the same afternoon, who gave details of a tender for which the firm was able to compete. By noon the salesman was given complete tender documents including drawings, prices, bills of materials and estimated duration of construction, and the order was obtained. In both these cases – which are merely examples of a commonly observed process – without the rapid response enabled by CAD, the firms would have failed to maintain their market presence.

6.3 Downstream benefits

In the earlier discussion in Chapter 2 we argued that the post-industrial revolution factory was structured into three distinct divisions – design, information control and production/distribution. In this form of factory organization information processing was paper-based and undertaken manually. Then, from the late 1950s, we began to see the penetration of various forms of digital, electronic processing of information in each of these three spheres of production. Nevertheless, despite this extensive penetration of computers, information flows between these divisions continued to be paper-based – moreover, in large part, that information which was generated within the enterprise (rather than coming from the external world, as in the case of market prices and conditions) was defined in the design stage. Consequently the move to the automated 'factory of the future'[8] has as its necessary pre-condition, the utilization of digital data generated during design and transferred directly to other spheres of the enterprise without 'unnecessary' human intervention.

At this time of writing it does not appear that this 'factory of the future' will emerge in a sudden puff of smoke in the Summer of 1982. It will inevitably be a process of *incremental*, although not necessarily slow, progress. Therefore the downstream links between design and information control, and design and production/distribution follow a similar pattern of gradual (but rapid) development.

Assessing the state of the art of these downstream benefits poses substantial difficulties, partly because of the difficulties in estimating the extent of these links, partly because they are gradual and emerging, and partly because of the very substantial variation in the nature of the potential links that depend upon the sector, the product and the type of factory organization. Therefore, once again it is only possible to *illustrate* the types of benefits deriving from the use of CAD as an example of the competitive advantages which utilization of the technology provides. For obvious reasons related to the earlier discussion we can distinguish between benefits related to information control and benefits related to production/distribution:

(a) *Information control.* The extent of information required by most firms which utilize CAD are generally beyond the processing capabilities of minicomputers and consequently in the case of most users visited, parts listings, bills of materials and ordering were undertaken by mainframe computers, separated from the CAD system. While it is quite possible for these different computers to inter-communicate directly, as a general rule there is some form of human interface between the two systems as well as some form of duplication of data generation and processing. However the more powerful mainframe computers have no such difficulty [9] and, as we have seen, this integration of data processing and analysis is a specific philosophy of IBM CAD system marketing. Moreover, as the turnkey CAD vendors move to the more powerful 32 bit computers (which they are all currently doing), their ability to integrate CAD-analysis, software and batch-information-processing will increase. Thus there is a growing tendency for the unification of data bases or for the intercommunication between different, distributed data bases. Once this is done, the degree to which firms will be able to reap substantial economies will depend upon their existing structuring of information. For example, one of the users visited (which designs and builds process plant) had already made substantial inroads into parts listing via their mainframe computer: as a consequence, on-site contingency costs had fallen from 15 per cent to 5 per cent. In at least three other cases, the pre-CAD organization of information was by contrast, sub-optimal. The installation of CAD systems in these enterprises had forced the downstream systematization of inventories. This led to a more structured organization of warehousing and numbering and (more importantly) in a reduction in the number of different parts held, as the systematization illustrated unnecessary proliferation of part-types. Downstream information flows need not be confined within enterprises, of course. A particularly pointed example emerges from the British motor component industry where a British automobile firm is building a new car under licence from a Japanese firm, and to its design. One of the major United Kingdom component suppliers complained that the specifications supplied by the Japanese firm for a particular co-ordinate were unintelligible – 'instead of sending us a drawing, all we got was a digital readout'! However, in general, most users reported that the intra-firm electronification of information flows was more feasible than that between firms. Another large British automobile firm and one of the United States aerospace firms reported that components were made by independent suppliers and consequently this inhibited the reaping of downstream benefits from their CAD systems.

(b) *Production/distribution*. There are very many potential links between design and production/distribution, and the specific benefits available depend upon the nature of the process and products involved. We illustrate some of these to give a flavour of some of the potential benefits which arise:

(i) Machine-setting. The control mechanisms of numerically controlled machine tools utilize the same basic digital information with which the CAD systems operate.[10] Consequently it is not a complicated task to link the two systems, and most of the firms interviewed had begun to establish links between CAD designs and NC milling and cutting machines. But in one case (which is a sign of a future trend) specifications for the automatic testing equipment were fed directly from the CAD system and in another case, active plans were being made to link assembly equipment to the CAD system. The benefit of these links are manifold, including the displacement of machine operators and a reduction in errors in machine setting.

(ii) Production planning. Once a unified (and accurate) data base has been established in the CAD system it is possible for this information to be assessed by multiple users. This not only spreads information more widely through the plant but many users reported that it speeds up the release of final drawings (in one case from three weeks to less than 8 hours) and allows for a single, corrected master design to be utilized, rather than the previously haphazard proliferation of incomplete drawings.

(iii) Materials saving. A particularly important benefit of CAD is a reduction in materials utilization due to the optimization of design and nesting (the name given to the programs developed to cut shapes out of a sheet of material). In one case, optimized designs had reduced the number of parts in a machine by 50 per cent; in another, CAD had made it possible to reduce silver utilization by 50 per cent in a process in which silver comprised 30 per cent of direct production costs. Benefits from CAD nesting programs are widely felt – one sheet metal user had reduced wastage from 40 per cent to 26 per cent in its first-generation nesting program; its annual savings in sheetmetal equalled the total annual wage bill.

(iv) Prototypes. Four users reported that as a primary benefit, more accurate CAD drawings reduced the need for, and costs of, manufacturing prototypes. In one particularly graphic case, an electronic instrument had been built for an aircraft (at a cost of $100,000) which did not fit into the space available in the cockpit! This required a complete re-design, which the firm argued would not have been required had a CAD system been utilized in the first place.

(v) Extra-firm benefits. A major and rapidly expanding field for CAD systems is 'piping' software which is used for 'interference checks' (i.e., whether pipes obstruct with each other or other sets of equipment in a process plant). Despite the fact that none of the existing turnkey CAD suppliers yet appear to have a fully mature 3D software package for this, [11] the benefits in construction are already being felt. In the past, a signficant but unquantified cost in the construction of process plant was incurred in the rectification on site of design errors which resulted in interferences. The elimination of such design errors is already beginning to have an impact in reducing construction costs, despite the immaturity of much of the available software.

6.4 Other relevant benefits from CAD

There are numerous other benefits arising from the use of CAD, of which two in particular stand out in relation to our interest in the impact of electronics in LDCs. The first concerns the ability which CAD gives to management to control proprietary information. In the case of the firm mentioned earlier (which was saved from extinction by the ability of CAD to produce drawings rapidly) the new product is not patentable. Consequently the full design is kept only within the CAD system to which only three people have access via an 80 digit code.[12] More specifically from the viewpoint of LDCs, one British heavy electrical manufacturer was able to take advantage of

CAD technology to the detriment of LDCs. In manual-drawing days, drawings provided to sub-contractors in developing countries frequently contained more information than was necessary and it was too expensive and time consuming to produce a specific set of drawings for each sub-contractor. Consequently, the firm observed, there was the inadvertent transfer of technology to LDC firms who subsequently became effective competitors. In this case CAD had given this British firm greater capability to control the amount of information contained in drawings passed to LDC sub-contractors and hence to limit the growth of potential competition.

A second area of benefit arises from the use of CAD as a marketing tool. A large number of users, in all sectors of production, remarked that their CAD systems were frequently used as a marketing tool – too frequently for most CAD system managers. Not only were potential customers shown around the CAD system – and often, while discussing the potential contract shown, through its use, the implications of changed designs – but a large number gave prominence in their annual reports to their CAD systems as an indication of the modern technology they were using.[13]

6.5 An overall assessment of the benefits of CAD

We have shown, with reference to the experience of users visited in the United Kingdom and the United States, some of the benefits derived from the use of CAD. The evidence suggests that (at DC equipment prices) at gross labour costs (including overheads) of over $11,000 per annum some firms are able to pay back CAD as a *design* tool within three years; when using low cost basic *draughting* systems on a double-shift basis, the switch-over gross wage costs are probably as low as $6,000 per annum. In addition there are a wide range of unquantifiable *upstream* benefits in design of products and *downstream* benefits in information control and production distribution.

These observations, based as they are on visits with users, do not necessarily imply that *all* firms can reap these benefits. Indeed at least two United Kingdom firms which were visited had clearly made very inefficient use of their CAD systems. These were not just a waste of money following acquisition, but were also a continuing drain on managerial resources. Moreover, the basis of sample selection in this study – predominantly, but not entirely, from CAD suppliers' users lists – almost certainly provides a biased sample in favour of efficient users since, naturally, most (but not all) suppliers were reluctant to put us in touch with 'failures'. It is difficult therefore, given the relatively small number of users visited, their nature and the immaturity of much of the CAD software involved (especially the low cost draughting aids), to develop rules of thumb which would ensure efficient utilization. However, two relevant factors stand out in determining the success with which CAD systems are utilized. The first concerns scale. Despite some exceptions[14] CAD design systems based upon minicomputers are unlikely to be profitable (in relation to the choice of technique, rather than the design of product) with fewer than ten designers; those based upon mainframes have even greater effective economies of scale, requiring around twenty potential users. And, secondly, for the real upstream and downstream benefits (i.e., excluding draughting) to be realized, the *system* of organization within which CAD is introduced is of absolutely critical importance for reasons which have been extensively discussed above. It is largely for this reason that, in our experience, the United States users made more efficient use of CAD – while the United Kingdom users, with weaker management and less experience, were often slightly tentative about their use of CAD, the American users tended to be more decisive. This difference is reflected in part by the differential CAD utilization rates between the United States and the United Kingdom.

In addition to these difficulties involved in successfully introducing CAD systems, there are a number of disadvantages associated with the use of CAD systems as a whole, whilst others reflect the problems which are faced by users pioneering any technology. One small scale American user had, in fact, anticipated such disadvantages prior to purchase, namely:

loss of only 1 man equivalent to 3 manual draughtsmen
impractical to fill in [absences and incomplete drawings] manually
susceptible to piracy [of manpower] as new installations made locally

The user had nevertheless gone ahead and installed a CAD system. So far, after 12 months of use, none of these disadvantages had been critical: in fact it was more guilty of pirating experienced draughtspersons from other firms than vulnerable to the loss of its own skilled staff.

To these disadvantages could be added the vulnerability of CAD to breakdown (especially with mainframe-based systems) and the danger of pirating designs that are stored centrally (One CAD systems manager observed that he could sell the master tapes of a non-patentable design for around $2 million). Nevertheless, despite these drawbacks CAD technology undoubtedly offers significant benefits to users. It is already mandatory for the aircraft, automobile and electronics sectors and we have every expectation that to compete in a wide variety of internationally traded goods and services, CAD will also become an essential requirement within the coming five to ten years. In the following chapter we attempt to specify which sectors will most benefit from the use of CAD technology.

Notes

1 For a description of this methodology and the same frame, see Appendix I.

2 CAD/manual productivity ratios in design are almost always lower than those in draughting.

3 Computervision claim that on average, their CAD systems are used for 88 hours per week in the United States, but less in Europe.

4 Despite the concern of trade unions, especially in the United Kingdom that flickering graphics screens would prove tiring to the eyes, none of the CAD operators interviewed felt this to be a problem.

5 The interest rates involved in the derivation of these costs were unfortunately not disclosed: it is assumed however, that they are at commercial rates. Moreover these equipment costs are at prevailing DC prices. In many LDCs (see Chapter 9) electronics equipment tends to cost between two and three times the prices prevailing in DCs.

6 Cheaper equipment is already available and prices are declining as microcomputer and data storage costs decline— the software component of these systems is minimal and static.

7 Although in general CAD systems require less working space than manual systems, particularly because they reduce the need for storing drawings.

8 The phrase used by General Electric in its emerging strategy to regain ground lost to more technologically advanced Japanese and European competitors.

9 Although when used together, CAD software requires priority over batch information-processing. In most cases batch-users do not even notice the consequences of this hierarchy of processing-access.

10 Although for this information to be utilized it has to pass through a post-processor which, in the absence of direct numerical control (DNC), prepares the paper tape for the computer-numerical control (CNC) machine tools.

11 As described before there is one package available from the United Kingdom (PDMS) which has recently been licensed by a number of large United States users.

12 In addition to the data tape within the CAD system, two other sets of the master tape exist – one in a fireproof vault in the design office and the second is stored in a bank.

13 One United States firm begins its annual report with a picture of its CAD system with the bold type quote from their General Manager:

> Our goal is to have a customer walk in with an idea in the morning and walk out with a prototype in his pocket the same evening.

The sub-type reads:

> . . . The General Manager . . . and Engineering Services supervisor, enthuse over the competitive advantages created by our new $400,000 Computer Aided Design (CAD) system. With CAD . . . can turn around engineering drawings and working prototypes almost instantly.

14 One particularly successful user employed only three designers.

7 THE DIFFUSION OF CAD TECHNOLOGY AND ITS IMPACT ON LDC COMPETITIVENESS

In the previous chapter we noted the very considerable benefits that CAD can provide to innovating firms, affecting *inter alia* the costs of production, the optimization of designs and the lead time with which products reach the market place. Yet if these remain potential, rather than realized benefits, then LDC enterprises are likely to be little affected. It is critical, therefore, to determine the rate at which this diffusion will occur and the sectors in which CAD is likely to diffuse. Only once this has been done will we be able to reach a view on the extent to which this technology will affect LDCs production for export, as well as production for their internal markets. These are the subject matters of this chapter.

7.1 The potential rate of diffusion

It is clearly impossible to know what the rate of diffusion of CAD technology will be since there are so many imponderables in the equation. It is, however, possible to make a reasoned guess. If Henwood (1980) is accurate in his assessment that penetration of United States industry in 1980 was 'less than 5 per cent' (say 4 per cent), and if CAD sales to manufacturing industry continue to grow at 40 per cent per annum, then assuming no growth in the absolute size of the United States industrial sector, penetration will be around 20 per cent by 1985 and around 100 per cent by 1990. These estimates are of course extremely crude, but they do suggest that under seemingly conservative assumptions, [1] penetration of DC manufacturing industry by CAD will become fairly widespread over the coming decade.

7.2 The likely path of diffusion

At one level every CAD supplier regards it of primary importance to determine the likely path of the diffusion of the technology. Yet at another level the suppliers have been so preoccupied with coping with the implications of compound growth rates of over 70 per cent per annum that little effort has apparently been put in this direction. Instead, generalized strategic decisions have been made (notably to stress the mechanical and ACE sectors) and established suppliers have concentrated on developing capabilities (such as numerical control applications programs) which will have an impact on a number of different economic sectors, rather than on systems specific to particular sub-sectors.[2]

At this broad level it is relatively easy to determine the path of diffusion. In the early and mid-1970s the electronics sector was clearly the dominant user of the technology. Penetration grew rapidly and almost all integrated circuit manufacturers in the West now use CAD as part of their standardized design procedures. But the penetration of CAD into downstream firms manufacturing and assembling printed circuit boards has been much more limited. This has in large part been due to the relatively high historic entry costs for CAD users, but the recent development of low cost dedicated microcomputer systems (see Chapters 3 and 5) has begun to significantly change this picture. Thus, unlike the integrated circuit industry where CAD is essential, utilization of CAD in most of the printed circuit board industry is largely a choice of technique problem and the technology has only recently become available at a low enough cost to justify widescale use. Nevertheless, despite this growing market for low cost systems, few established CAD suppliers regard the electronics sector as a leading sector of diffusion in the future.

From about 1978, especially in the United States and Japan, the mechanical engineering market began to 'take off' so that most large-scale CAD suppliers regard it as now being the primary market. Not only is it the largest area of current applications (see Table 4.3) but the absolute size of the mechanical engineering sector makes it the largest potential user as well. One source (Henwood, 1980) estimates penetration in the United States manufacturing sector (the most advanced user of all countries with the possible exception of Japan) as being less than 5 per cent, suggesting plenty of scope for continued expansion. But despite this rapid growth of the mechanical engineering market, the consensus in the industry appears to be that the major sectors for expansion in the second part of this decade will be the architecture and civil engineering sectors. Here CAD technology is currently only used in the mapping fields and in pipe-work design, but potential applications in other areas of process plant design, engineering works and building design are enormous, particularly in small-scale architectural and civil engineering practices.

From the point of view of LDCs these generalized observations on sectoral diffusion are only of limited value. It is necessary to have a much more detailed picture of the path of diffusion if sensible policies are to be drawn up. But assessing this more detailed pattern of diffusion creates enormous difficulties and the most we can hope for, in the analysis which follows, is to establish some guidelines for evaluation. Our procedure will be to refer back to the analysis in Chapter 6 on the benefits arising from the use of CAD. In this we classified three major areas of benefit, namely in draughting, in product design, and in downstream activities. We thus hope to generate a perspective on the likely paths of diffusion by examining the economic sectors affected most significantly by each of these three areas of benefits:

(a) *Diffusion through benefits in draughting*. We observed in the previous chapter that at gross labour costs (that is, including overheads) of around $6,000 per annum CAD can be justified as a draughting tool if the system works on a two-shift basis. This is a pure choice-of-technique problem and given the existing costs of employing draughtspersons in all DCs and a great number of semi-developed LDCs, [3] it clearly makes economic sense for many firms to use CAD technology for draughting. By so doing firms will be able to reduce their design costs, although the relative significance of these savings will vary between sectors and over time. One indication of these relative draughting costs is the proportion of draughtspersons in different economic sectors is difficult to obtain. The only comprehensive information available is that for the United States in 1960.[4] While in some senses the industrial structure of the United States in 1960 makes an appropriate comparison for LDCs which tend to use relatively dated technology, the datedness of this data has two severe drawbacks. Firstly we are primarily looking at the economic impact on DC firms. And secondly, the most significant change in industry structure (rather than in technology within any particular industry) since 1960 has been the development of the electronics, scientific instruments and office machinery branches. These are barely represented in the data for 1960.

Given the use of CAD as a draughting tool and given that the technology represents an optimal choice of technique in DCs, we can expect that it will diffuse through sectors in relation to the proportion of draughtspersons in their labour force. If this is the case then it can be seen from Table 7.1 that the technology is likely to go first to those higher-technology sectors in which LDCs are trying to develop specializations (e.g., electrical equipment, office equipment, machinery and shipbuilding) and then only latterly to those sectors in which LDCs traditionally have a comparative advantage (e.g., agricultural products, fisheries, leather, apparel, textiles and food products).[5]

Table 7.1 Draughtspersons as a proportion of the labour force: United States in 1960 (per cent) †

	High intensity			Low intensity	
	Electrical equipment	2.08		Agriculture	0
*	Office equipment	2.00		Fisheries	0
	Professional and scientific equipment	1.94	*	Leather	0
	Machinery except electrical	1.92	*	Leather footwear	0
*	Railroad equipment	1.82		Leather products	0.01
*	Shipbuilding	1.77		Apparel	0.02
	Fabricated metal products	1.53		Tobacco products	0.04
*	Petroleum and gas extraction	1.53		Textile mill products	0.04
*	Aircraft	1.37		Miscellaneous manf's	0.05
	Durable manufactures	1.21	*	Canning, preserving, freezing	0.05
	Transport equipment	1.18		Food products	0.06
*	Farm machinery and equipment	0.97		Printing and publishing	0.12
	Manufacturing – total	0.74	*	Coal mining	0.14
*	Motor vehicles	0.74		Non-durable manf's	0.15
	Construction	0.65	*	Cement products	0.15
*	Furniture and fixtures	0.48	*	Non-metallic mining	0.18
	Petroleum refining and coal products	0.48		Wood and wood products	0.23
	Chemicals and allied products	0.46		Stone, clay, glass products	0.24
	Primary metal products	0.38	*	Synthetic fibres	0.27
*	Paper containers	0.36	*	Drugs and medicines	0.27
*	Blast furnace and steel products	0.34		Plastic products	0.30
	Forestry	0.34	*	Metal mining	0.30
	Rubber products	0.33	*	Pulp and paper	0.31

Median Paper and allied products 0.32.

† Based on ISIC two-digit classification, except for those sub-sectors marked with an * which are at the three-digit level and are therefore sub-divisions of two digit branches already represented in this table.

Source: Calculated from the United States Department of Labour (1969).

(b) *Diffusion of CAD as a design tool.* In determining the diffusion of CAD as a design tool we need to make an important distinction between design intensity and design sensitivity. In the former case CAD can once again be seen as a choice of technique (i.e., to reduce design costs by savings on design skills). Whereas the sensitivity of a product to a particular, strategic design input need bear no necessary relationship to the product's design intensity:

(i) *Design intensity*. Table 7.2 is drawn up on a similar basis to Table 7.1 except that it reflects the proportion of engineers, kindred technicians, and architects in the labour force. It has the same drawbacks as the data for draughtspersons, particularly in relation to the electronics-related sectors. But to the extent that design intensity is an indicator of the path of diffusion, then we can expect CAD technology to penetrate economic sectors in relation to the ratios expressed in Table 7.2. Once again the sectors in which LDCs have traditionally specialized have a very low design intensity. By contrast those sectors in which LDCs are hoping to specialize in the future have a high design intensity [6] – in the case of the electrical machinery sector, for example, around one in six employees is classified as a designer.

(ii) *Design sensitivity*. Design sensitivity need not correspond to design intensity. For example the design of shoes is probably the most critical component of market growth, yet this sector has a very low design intensity. Similarly, an American firm has recently set up a United Kingdom plant to sand-blast designs, via CAD, onto glass tumblers. The design intensity of this sector is low, yet the sensitivity to design changes (which in this case are facilitated by CAD) are very high. In assessing the design sensitivity of any sector, there are no indicators to help us in establishing the relative ordering of sectors. In some cases (e.g., VLSI chips), the conclusion is unambiguous – the product could not be made without CAD. Included in this category are most computers, ULA,[7] LSI and VLSI chips, nuclear power plants and civil aircraft. But in many other sectors, although CAD is not a necessary condition for design, optimization through the use of CAD is a very significant determinant of competitiveness. This is reflected not just in function (e.g., the drag coefficient of motor cars) but also in the speed with which the products reach the market place. Therefore we have little way of judging the trend except to observe that, in general, intermediate products are often fairly standard and are not characterized by differentiation for consumer markets. Consequently, they are not generally likely to be design sensitive, although they may well be design and/or draughting intensive and therefore be 'ripe' for the early introduction of CAD.

(c) *The diffusion of CAD through its impact upon downstream activities*. As with design sensitivity there are no indicators to enable us to assess the future path of diffusion as a consequence of the downstream benefits arising from the use of CAD. It would be

Table 7.2 Engineers, kindred technicians and architects as a proportion of the labour force: United States in 1960 (per cent) †

	High intensity			Low intensity	
	Electrical machinery	17.8		Apparel	0.08
	Transport equipment	12.22	*	Leather shoes	0.09
*	Aircraft	10.67		Printing and publishing	0.12
	Machinery except electrical	10.34		Fisheries	0.13
	Professional and scientific equipment	8.13		Leather products	0.15
*	Office equipment	8.07		Wood and wood products	0.17
	Non-durable manufactures	5.57		Tobacco products	0.17
	Petroleum refining	4.72		Textile mill products	0.24
*	Synthetic fibres	4.57		Food products	0.33
*	Petroleum and gas extraction	4.40	*	Leather	0.33
	Durable manufacturers	4.26	*	Canning, preserving, freezing	0.38
	Fabricated metal products	4.21		Agriculture	0.61
	Chemicals and allied products	4	*	Coal mining	0.73
	Farm machinery and equipment	3.88	*	Cement products	0.97
*	Shipbuilding	2.96	*	Paper containers	1.28
	Miscellaneous manufacturing	2.89		Stone, clay, glass products	1.31
	Manufacturing – total	2.80		Rubber products	1.39
*	Railroad equipment	2.62		Plastic products	1.40
*	Metal mining	2.55	*	Non-metallic mining	1.42
	Primary metal products	2.28	*	Drugs and medicines	1.43
	Motor vehicles	2.24		Forestry	1.52
*	Blast furnace and steel	2.18		Paper and allied products	1.58
	Construction	2.15	*	Furniture and fixtures	1.82

Median * Pulp and paper 1.89.

† Based on ISIC two-digit classification, except for those sub-sectors marked with an * which are at the three-digit level and are therefore sub-divisions of the two-digit branches already represented in this table.

Source: Calculated from United States Department of Labour (1969).

convenient for analysis if this effect was neutral between sectors, but this is almost certainly not the case. The construction industry, for example, with its heavy reliance on parts lists, bills of materials and numerous sub-contractors, seems to benefit particularly significantly in this regard. But the benefits that arise from heavy component and materials use are only one way in which CAD technology has a beneficial impact upon downstream costs and our conclusions with regard to sectoral diffusion must therefore be agnostic.

Finally, in a recent study on the manpower implications of CAD in the United Kingdom, Arnold and Senker (1982) have also made an attempt to gauge the diffusion of the technology. They determine two characteristics of industries – 'high technology' corresponding to our design intensity, and 'high draughting' corresponding to our draughting intensity. Using restricted United Kingdom government data on the distribution of manpower, they relate the known utilization of CAD to individual sectors. The category 'high technology, low draughting' ranks highest in the relative use of CAD, followed by 'high technology, high draughting', 'low technology, high draughting', and 'low technology, low draughting'. This confirms our earlier observations, based on the 1960 data from the United States, that the degree of design and draughting intensity provide an accurate indication of the sectors in which CAD technology is likely to diffuse.

7.3 The impact of CAD upon LDC exports of manufacturers

CAD is only one of a bundle of advanced electronics-based technologies that are diffusing through DC manufacturing industry. It would therefore be wrong to place too great an emphasis on the potential impact of CAD alone on LDC exports of manufactures. But given that its use does confer substantial benefits upon the innovating firms (see Chapter 6) and given that it is representative in many ways of other electronics-related technologies (Chapter 9) it is clearly of consequence to relate the potential path of the technology's diffusion to the future growth of LDC exports.

In Table 7.3 we show the extent of growth of all imports of manufactures by DCs from LDCs between 1970 and 1978. We also relate the ranking of these manufactured exports, by size and by growth, to the earlier information generated on design and draughting intensity (Tables 7.1 and 7.2). A number of relevant observations can be drawn from this table. Firstly the growth in overall manufactured exports (73 per cent) over the eight years was remarkably high. Secondly, exports of higher-technology manufactures grew more rapidly than those of traditional manufactures, although the difference between these two groupings was not as large as some observers suggest (e.g., Lall, 1979). Thirdly, textiles and garments remained the largest group, alone accounting for 48 per cent of all DC imports of manufacturers from LDCs in 1978 (but down from 55 per cent of the total in 1970). Finally there is a very pronounced tendency for traditional manufacturers to be of both low intensity and low draughting intensity – by ranking they fill the last four places in each case.

Table 7.3 DC imports of manufactures from LDCs in relation to design and draughting intensity

	Value $ million		Growth	Rankings (N=15)			
	1970	1978	1978/1970	Value (1978)	Growth	Draughting intensity	Design intensity
Traditional manufactures							
Semi-finished textiles	1,815	9,610	5.3	1	13	11	11
Leather	183	950	5.2	9	14	13	12
Clothing	1,181	9,502	8.1	2	10	12	14
Shoes	151	2,033	13.5	7	6	14	13
Higher-technology manufactures							
Chemicals	588	2,282	3.9	5	15	9	6
Metals and metal products	319	2,223	7	6	12	10	9
Machinery except electrical and business	81	1,136	14	8	5	4	3
(Farm machinery)	2	29	14.5	15	4	7	7
Electrical machinery	372	4,463	12	3	7	1	1
Business machines	81	600	7.4	12	11	2	5
Scientific instruments	24	359	15	13	3	3	4
Motor vehicles	23	603	26.2	11	2	8	10
Aircraft	18	737	40.9	10	1	6	2
Shipbuilding	40	355	8.9	14	9	5	8
Consumer electronics	214	2,391	11.2	4	8	*	*
Total manufactures	5,493	40,195	7.3				
Total traditional manufactures	3,330	22,095	6.6				
Total higher technology manufactures	2,163	18,100	8.4				

* Datedness of 1960 data (see Tables 7.1 and 7.2) does not allow for meaningful figures. In general design is high in this sector and so is draughting.
Source: Calculated from United States Department of Labour (1979) which provides information on ISIC sectors, and United States Central Intelligence Agency (1980), which provides information on Standard International Trade Classification (SITC) sectors.

We argued earlier that we can expect that CAD will diffuse first to those sectors that have relatively high design and draughting intensities. In so doing they will cut the costs of design, which in some cases are substantial. In the electrical machinery sector, for example, one in six of the labour force are in the design category; in process plant design costs can be as high as 20 per cent of total fixed capital costs, and up to 50 per cent in the case of nuclear power plants. The gains in the productivity of labour in design and draughting of around 3:1 which CAD enables can therefore be of considerable significance in reducing production costs of innovating firms.

Relating this argument to the information presented in Table 7.3, it is clear that the sectors in which CAD will have the greatest impact in reducing design costs are precisely those higher-technology sectors in which LDC manufactured exports grew most rapidly in the 1970s and in which many of them hope to specialize in the 1980s. Moreover, a further set of rapidly growing high-technology exports from LDCs (especially to other LDCs) which is not represented in Table 4.3, is in the ACE sectors (O'Brien, 1981). These sectors are relatively high in both draughting and design intensity, and their related activities in the construction sector are also particularly high in their draughting intensity. With the increasingly rapid build-up of relevant CAD applications programs we can anticipate that the technology will diffuse rapidly in these sectors.

However, as we have already observed, the reduction in design and draughting costs is only one of the benefits that arise from the use of CAD. The two other sets of benefits are product optimization and the reduction of downstream costs. The problem is that we have no systematic way of gauging how these two sets of benefits will affect the diffusion of CAD through different sectors. There is no justification for assuming that in the future it will be systematically related to either design/draughting intensity or to traditional/higher technology sectors as it appears to have been in the past. For example, with regard to the garments sector which accounted for 24 per cent of all DC imports of manufacturers from LDCs in 1978 (up from 21.5 per cent of the total in 1970), and which has a low design/draughting intensity, we know that the primary impact on production costs of electronics related innovations has been in the use of CAD cutting and nesting techniques (Hoffman and Rush, 1981). Similarly, although food products are neither design nor draughting intensive (see Tables 7.1 and 7.2), their packaging frequently is. Moreover sales of these commodities in DC markets are often highly sensitive to design changes. But we also know, however, that a number of non-traditional manufactures (e.g., electronic toys and integrated circuits) are particularly sensitive to changes in design (including the speed with which products are launched).[8]

In summary, therefore, we have argued that CAD is likely to diffuse rapidly through DC industry in the coming decade. Furthermore, it is probably that the path of diffusion will reflect the benefit derived from the use of CAD. Focusing on the export performance of LDCs over the past decade, we can conclude that CAD is therefore likely to diffuse first to precisely these non-traditional sectors in which exports grew most rapidly during 1970–1978 and in which many LDCs are hoping to specialize in the 1980s. Consequently, unless LDCs enterprises are able to introduce CAD in these sectors, they are likely to display diminished competitive capabilities in the market place and be forced back to the export of traditional manufactures, to the extent that these are less signif' antly affected by CAD technology.

Notes

1 The real growth of sales to industry in the United States has been well in excess of 40 per cent per annum since 1978 (see Chapter 4). However the assumption of zero growth of United States manufacturing industry over the coming decade is probably too restrictive.

2 By contrast, as we saw in Chapter 4, some small-scale suppliers producing dedicated microcomputer driven systems have confined themselves to particular industrial subsectors (e.g., Grafcon producing CAD for small-scale building design discussed in Chapter 4).

3 Although the acquisition costs of CAD equipment in LDCs are almost certainly higher than those in DCs.

4 At a broader level of aggregation more recent data is available for the United Kingdom (Fidgett, 1979). In 1978 the proportion of draughtspersons as a percentage of the labour force was 3.12 per cent in mechanical engineering, 1.7 per cent in instrument engineering, 1.85 per cent in electronic engineering, 1.53 per cent in vehicle manufacture, 1.12 per cent in metal goods sector, 4.1 per cent in marine engineering and an average of 2.04 per cent in all engineering sectors. It is not clear from this study by Fidgett why the proportions of draughtspersons in the United Kingdom are so much higher than in the United States (see Table 7.1). We presume it follows from a differential definition of 'labour force' between the two studies, but since it is the rank ordering which primarily interests us, the divergent proportions are of not too great a concern.

5 For a more detailed discussion see Table 7.3 and accompanying discussions later in this chapter.

6 However despite the fact that we reached a similar conclusion with regard to draughting intensity, there is no detailed correspondence between the categories in these two tables. The correlation coefficient (R=0.04) is insignificant.

7 A new and very important potential trend in the electronics sectors concerns the emergence of uncommitted logic array (ULA) chips. Unlike existing chips, the final logic of the ULA chip is decided by the user, rather than the manufacturer. The subsequent connection of 'uncommitted' logic gates *cannot* be done without the use of a very sophisticated CAD facility. According to one knowledgeable source (Dr I Mackintosh of Mackintosh Consultants) there is at least an even change of ULA chips becoming dominant with the next five to ten years.

8 This has been the primary factor that has led the motor vehicle industry (which is not particularly design intensive – see Table 7.2) to adopt CAD so vigorously. It is now one of the industries most significantly penetrated by CAD.

8 SKILL REQUIREMENTS FOR OPTIMAL USE OF CAD

In an increasing number of sectors, therefore, CAD either already is, or is rapidly becoming, a mandatory technology if firms are to compete effectively in growingly competitive markets. Our attention is thus naturally drawn towards the skills required to make efficient use of this radical new technology. Earlier discussion pointed out that, as with many new electronic technologies, the gains derived from using CAD are not only felt within the narrow confines of draughting and design but span the whole system of factory organization. Consequently in discussing the skills involved in its efficient utilization (which is the subject matter of this chapter), we must consider both operator- and management-skills, as well as the learning curves which are involved. At the same time a software-intensive technology such as CAD requires extensive back-up and service skills from the suppliers if it is to run efficiently.

8.1 Skills

8.1.1 Operator skills

There was a fairly strong consensus amongst all users visited (bar one) [1] that, at the worst, the skills required to operate CAD systems are the same as those involved in manual systems, and in more favourable circumstances, the required skills are lower than those with which traditional draughtspersons are equipped. As far as *draughting* is concerned, CAD (like other forms of mechanization) de-skills the job and reduces it to machine operation. By contrast in the case of *analysis and design*, by removing the unskilled element, CAD distills the skill component for concentrated attention by the designer. We see here a difference in the position of draughtspersons and designers – the skills of the draughtsperson are, in effect, boring constraints for the designer.[2]

The reason why both operators and management concluded that CAD reduces the skill component in draughting is because it removes the craft element (requiring extensive practice) associated with individually tailored layouts and individually developed lettering. This is not to say that CAD has no specific skill requirements but the type and levels of required skills are altered. Most management reported that typing skills were an advantage, given that all CAD systems have an alpha-numeric keyboard for inputting at least some of the data. But, as far as management is concerned, attitude and flexibility are more important attributes; some management felt (with no apparent justification) that these were a function of youth;

one system manager claimed that 'females were less flexible and dexterous' despite the existence of many female CAD operators, particularly in the United States. Finally IBM, which has a particularly rapid screen response time, argues that manual dexterity is an important attribute – thus while the average IBM CAD operator works at around 30 interrupts per minute, the best United States performance is 100 interrupts per minute. (Although, clearly, the level of response will depend upon the nature of the work.) Thus it was the general view of most management that although a good draughtsperson makes a good CAD operator, the CAD system was sometimes able to make a good operator out of bad draughtsperson (e.g., someone who was sloppy at lettering).

8.1.2 Management skills

The skills required by managers (as opposed to operators) of the CAD system are of a more substantive nature, for three reasons. First, as we saw in the previous section, many of the benefits of CAD are felt downstream. The current state of the art of CAD technology is generally one in which the basic draughting gains have been mastered and the cutting edge of progress lies in capturing these downstream benefits. While this latter task is partly a function of improved application software, it also requires a changed attitude by management, which is still feeling its way towards new forms of system's organization. Second, even within the draughting/design stage, management has an important role to play in improving productivity. Whilst operators are able to build up their skills until they can use the CAD system rapidly, for the real benefits to be obtained the draughting offices have to be well organized and 'menus' [3] have to be built up. For example, as we saw earlier, one United Kingdom machinery manufacturer obtains most of its productivity increase out of modular designs in which standard drawings of sub-assemblies are stored within the CAD system's memory and can be pulled out by the designer. In another case, the benefits realized in printed circuit board layout depended upon a re-organization of office precedures – in the earlier time period, basic layout drawings were made in any form, whereas optimal CAD layout depended upon a uniform presentation of these basic drawings. And finally, management argues that it has an important function to perform in evening-out the performance of individual operators into a curve of continuous and harmonized growth for the CAD system as a whole.

Given these taxing demands upon management, it is not clear

that any particular skill (different, that is, to pre-CAD skills) is required. However a *vision* of systems gains and the *power* to implement organizational changes, are of critical importance. And to the extent that these attributes are important it is not evident whether these can be formally learnt or whether they are acquired through direct experience.

It appears as if the ability to implement CAD systems efficiently is largely a function of the managerial level with which CAD is introduced. While all CAD suppliers aim to have CAD backed at the highest level, few have the 'clout' of IBM to enforce changes on its users. As a general policy, IBM insists that:

(i) before installation they meet with senior management to discuss the systems changes that are involved;
(ii) the best managerial manpower should be devoted to the system; and
(iii) the decisions on required work-changes (e.g., on the procedure for releasing final drawings) are worked out in detail before installation.

The less efficient systems we observed amongst users were generally associated with weak management and/or management with inappropriate training. More specifically, it appears that CAD systems placed in the hands of management trained primarily in data processing were used too narrowly to take advantage of systems gains.

8.1.3 Back-up services

CAD systems are new, different and software intensive. Users consequently require a great deal of back-up from suppliers until their systems function efficiently, particularly when the innovating firm is not experienced in using analagous types of equipment (e.g., automatic testing equipment). This ability to offer assistance to users is, as we saw in Chapter 5, an important competitive strength in the supplying industry. All turnkey manufacturers (bar the suppliers of small dedicated systems) are rapidly expanding their servicing centres. As a rule of thumb, for example, Computervision aim to have in Europe one application engineer (providing software support) for every eight systems in use, and one maintenance engineer (for hardware) for every four systems. In addition there are a further twenty-four software staff in the United Kingdom alone who service the home market and parts of the European market.

The services offered by the suppliers relate predominantly to software. Most suppliers reported that over 50 per cent of software 'bugs' turned out, on inspection, to reflect the 'ignorance' of users. But in addition to the on-going services which suppliers offer, the growth of users' systems efficiency is enhanced by specialized courses which all suppliers provide (usually at a price) as well as by regular user groups in which users can exchange experience, exchange (or market) software programs they have written and collectively place pressure on the suppliers to improve their software and support. The dominant conclusion that emerges from all this is that 'closeness' to CAD suppliers is important for efficient use, requiring an on-going, 'handholding' relationship. Similarly, closeness to other users is also a valuable asset.

8.2 Learning curve

Obviously the learning curve of both operators and management largely depends upon the use to which the CAD system is put. For example a system used for pure draughting is generally easier to assimilate than one used for design or the capturing of downstream benefits. On the other hand, such limited uses (characteristic of unimaginatively managed systems) may well reflect the inability of management to learn to use CAD effectively.

As a general observation, it appears as if there is a trade off between the rate of learning (of both operators and managers) and the ultimate success with which the system is used. The too-rapid definition of procedures that assist in obtaining early gains in productivity, has the effect of boxing-in the future potential of the system. In one United States firm visited, two CAD systems had been purchased and given to largely autonomous divisions making different sorts of integrated circuits. One had gone for rapid productivity gains and appeared to have a steep learning curve while the other had been deliberately used in a more reflective, exploratory way. The firm, after nine months, was beginning to assess the degree to which this trade off between speed and capability affected their ultimate designs.

The slope of the learning curves can be steepened in a number of different ways. One is to send both operators and managers off on courses, or to institute formal in-house training programmes. In fact most CAD suppliers included a series of basic courses in their sales price and offered a range of specialized courses for particular applications programs. A second strategy, generally pursued by enterprises for whom rapid progress is vital, is to poach expertise from existing CAD users. This presents gains to both the purchasing firm and the individuals involved.[4] Third, management often gained important experience from conversations with counterparts at the users' groups organized by most turnkey suppliers.[5]

On the basis of discussions with both management and operators in firms using CAD, it is possible to simulate a series of learning curves, as is done in Figure 8.1 below. Insofar as operators are concerned learning occurs in a series of steps. At around three months, or earlier, most operators are up to the level of productivity of manual systems. There is then a short period of retrenchment while the operator assimilates the newly learnt basic operating skills that are thereafter continually improved via the assistance of menus, until around six months. Then the operator begins to take advantage of the sub-designs stored in the CAD system and makes more effective use of analytical programs. The majority of users reported that by the end of the first year, most operators were about as efficient with the system as they could ever get – further benefits depended upon the back up provided by systems organization. There was, however, some difference in the views of users about whether the period of individual retrenchment was associated with a flattening of the curve or a temporary decline in productivity – since we encountered sufficient of these latter views we believe that a temporary fall-off in productivity is a common experience.

One of the functions of management is to iron out these periods of individual retrenchment and thereby to increase the general productivity of the system by moving the individual operators to a more desirable learning curve. This can be done by providing special assistance to operators experiencing difficulties, but it also sometimes appears to involve the slowing down of particular rapid learners who, by pulling ahead of their colleagues, have the effect of imbalancing the system [6] – see Figure 8.1 But management also has a variety of other functions (described above) involving the generation of menus and the establishment of procedures which affect the learning curve of particular operators.

But ultimately, as we have repeatedly pointed out, the productivity of the CAD system depends upon organizational changes in the system, including complementary adjustments in the other spheres of the firm (i.e., information control and production/distribution). Without these, the full benefits of the CAD system cannot be realized. The time span within which management is able to comprehend the changes that are necessary, and then to implement them, is generally – but not always – longer than those involved in the individual learning curve. However there are, as with the operator learning curve, periods of flatness in the growth of system productivity.

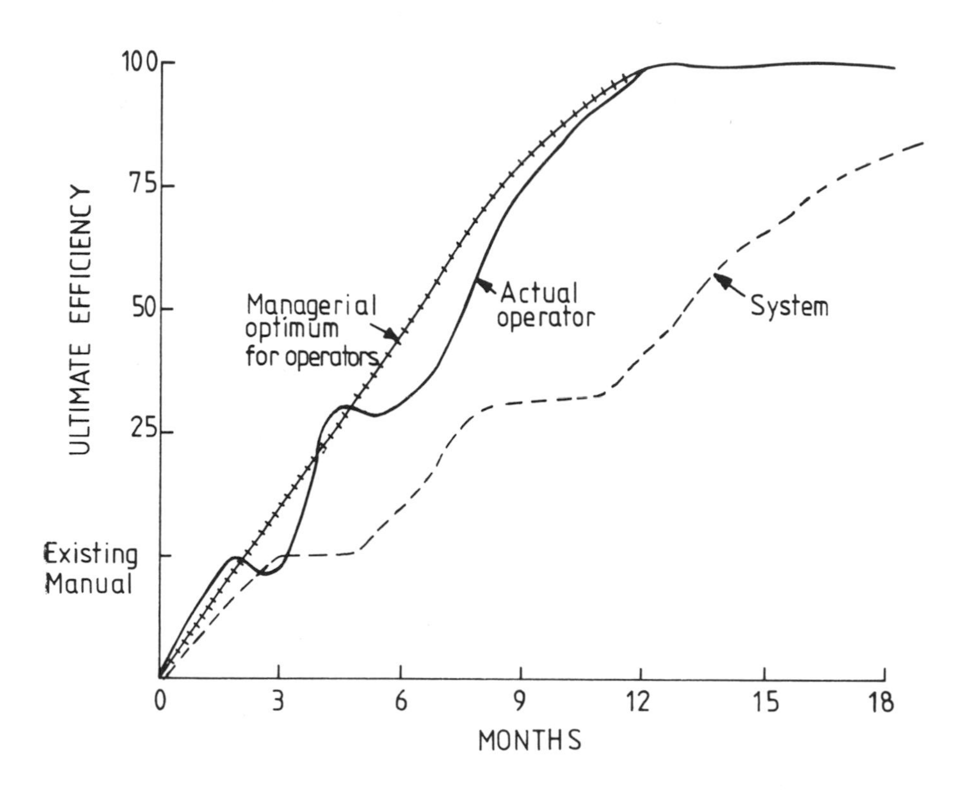

Figure 8.1 Hypothetical learning curves

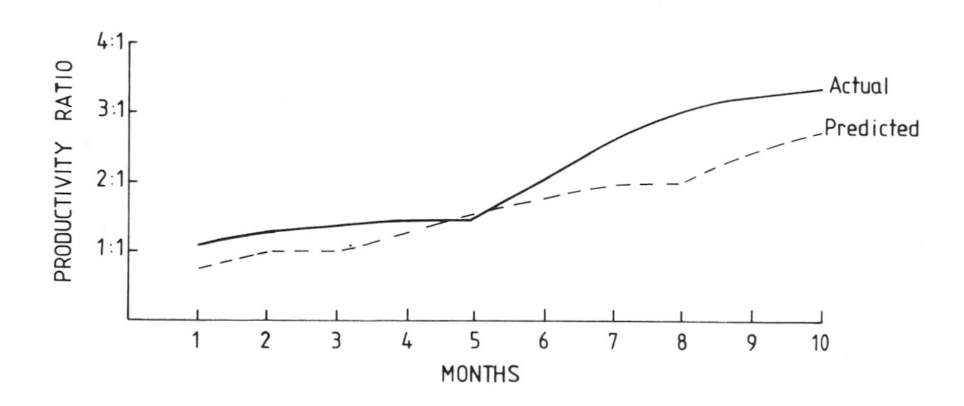

Figure 8.2 Learning curve – system.
(Source: United States user (electronics application)

Evidence for this is given in Figure 8.2, based upon the actual (and predicted) learning curves of a United States manufacturing firm using CAD in predominantly electronics applications. After one month the system was equalling that of the manual system, but progress thereafter was slow until five months, when complementary organizational changes were implemented, after which progress was rapid. By ten months the productivity increase was 3.3:1 (compared with a predicted 2.8:1), and, characteristically, the firm was unsure at this stage how far up the ultimate learning curve it had progressed.

Notes

1 The dissenting firm used CAD primarily for design, and in our view was confusing a general improvement in the quality of their design staff with the effect of the CAD system.

2 This differential impact upon draughtspersons and designers is reflected in the attitude of trade unions to the introduction of CAD. In the United Kingdom the draughtspersons' trade union (TASS) has been most active in resisting the diffusion of CAD, while the designers' union (ASTMS) has in general had no objection to its introduction.

3 See Chapter 3 for a description of menus.

4 This mobility of manpower is more extreme in the United States than Europe. But American turnkey suppliers are rapidly spreading the message. The British head of one United States supplier was formerly in charge of CAD evaluation in one of the largest United Kingdom users; another supplier had recently poached a CAD operator (of a rival system) and given him a car and a salary hike of over 30 per cent.

5 These users' groups have the added advantage of unifying pressure on suppliers to improve their service and de-bug their software.

6 In one firm a graduate was employed as an operator, learnt how to use it in four days and then left as he felt constrained by the system.

PART III

THE IMPACT OF ELECTRONICS ON THE INTERNATIONAL DIVISION OF LABOUR

9 THE POLICY IMPLICATIONS

9.1 An assessment of the overall impact of electronics

At various stages in this book we have drawn on perspectives generated by other studies on the role of electronics in economic performance, set up hypotheses in the potential impact of electronics on LDCs, and reached conclusions with regard to the impact of CAD technology. The stage is now set for an assessment of the wider issues involved, and for the prescription of policies for LDCs to enable them to adapt successfully to the rapidly changing economic and technological environment. But, before doing so, it is first necessary to retrace our steps over some of the ground that has been covered in earlier analyses.

In Part I we set the context for this study in relation to long-run cycles of economic activity. Crucially, we drew upon two conclusions that have emerged from the work of contemporary economic historians. The first was the recognition of the 'heartland' nature of electronics-related technologies, which represent not just one amongst a variety of incremental innovations, but a family of key technological developments which are in large part intimately related to the current 'long wave' of economic activity. And secondly, we observed that each of the long waves are characterized by an expansionary upswing, followed by a rationalizing downswing when the heartland technologies are used to meet growing competitive pressures. LDCs prospered (i.e., in the growth of their industrial sectors and in terms of their ability to export manufactures) in the upswing – the major object of this book is to throw some light on how they will fare in the emerging downswing, particularly as electronics-related technologies diffuse downstream into other sectors.

CAD technology is unusual in that its development represents both the upswing and the downswing of the current cycle. In its earlier phase it was a new product, facilitating the development of the electronics sector itself. But now, in the 'rationalizing' stage of the long-wave cycle, it has begun to diffuse downstream into industry, enhancing the ability of innovating firms to cope with markets that have become growingly competitive, at least in part due to the efforts of LDC producers. Moreover, because of its role in defining the data base for enterprises, CAD technology is a crucial stepping stone in the progress to the automated 'factory of the future'.

Chapter 6 was devoted to an analysis of the benefits arising from the utilization of CAD technology. In it we noted that some of these benefits were inherently difficult to measure. We also noted that the degree of benefits actually realized depended not only upon the products and processes involved, but also upon the extent to which systems-gains were captured. Thus, because of the wide variability in the use of the technology, and because of the difficulties associated with the re-organization of systems of production, it is inevitable that there has been a wide range in the benefits realized by enterprises using CAD technology.

Nevertheless, despite this variation, the benefits arising to firms that use the technology have proved to be substantial. Broadly we can classify these into five major areas, namely:

(a) In *product*, where without CAD the design of some products (e.g., integrated circuits) would be impossible, and that of other products (e.g., cars, process plant) would be sub-optimal.
(b) In *process*, where CAD (used on a double-shift basis) is an optimal draughting technique at wage costs (including overheads) of around $6,000 per annum and, in design, at annual wage costs of around $11,000.
(c) In *lead-time*, where without the use of CAD, products would reach the final market place substantially later, often involving a considerable competitive disadvantage to firms which make no use of CAD.
(d) In *marketing*, where the use of CAD not only provides the image of modernity, but allows the design to be easily optimized in a 'real time' interaction between the customer and the supplier of a product.
(e) In *control*, where the centralization of the data base in a CAD system allows for more effective appropriation of technology, and where its greater flexibility enables more selective dissemination of information to sub-contractors.

Therefore, to the extent that CAD is representative of the sorts of benefits arising from the use of electronics-related technologies in other sectors, we can make some tentative generalization on the impact of electronics on the ability of LDCs to compete in DC markets.[1] With respect to the choice of technique, it is clear that many LDCs may be able to do without the latest electronics technologies. It is true that they do significantly increase the productivity of labour (by around three times, in the case of design and draughting), but this does not necessarily imply that the older technologies are economically inefficient.[2] Insofar as systems gains are concerned, the situation is less clear. CAD, and by inference other sorts of electronics technologies, clearly have a synergistic relationship with complementary electronics technologies. But because of the difficulties involved in measuring these external economies, and because many of them will only be realized over time, it is difficult to assess their impact of LDCs. But, if CAD is a representative example, the primary impact of electronics related technologies will be in the sphere of product development and marketing, with the flexibility of the technology allowing for substantial gains in the speed with which new and optimized products can be introduced.[3] Moreover, the enhanced control over technology and the marketing advantage that arises from the use of CAD lend additional support to the conclusion that the introduction of electronics technologies such as CAD will be mandatory if LDCs are to continue to compete in DC markets, particularly with regard to non-traditional manufactures. Indeed, it is precisely in these non-traditional sectors (where LDC export growth has been most rapid over the past decade, where value-added is highest and where LDCs are hoping to specialize) that CAD technology seems most likely to penetrate first.

However, the evidence so far suggests that CAD technology, and by inference other types of electronics-related innovations, has been very slow to diffuse to LDCs. Of over 8,000 systems sold in the major sector of the industry (which has been analyzed), only thirty-two have gone to LDCs and many of these have been in non-manufacturing sectors. Of the LDCs only India appears to have an embryonic capability in producing CAD systems, but these are immature and have barely penetrated through to industry.

The portents for LDCs, therefore, are bleak. They are faced with a family of rapidly diffusing electronics-related innovations, which provide significant benefits to innovating firms – particularly with regard to the quality of products and their speed of introduction. Yet not only are LDCs slow to use these technologies but CAD suppliers report that LDCs are largely ignorant of their existence, or the capability of their products. It is tempting therefore to conclude that the technological gap between LDCs and DCs, which has appeared to close since World War II, is once again beginning to open. Moreover, insofar as DC enterprises are introducing these electronics-related technologies more quickly than LDC counterparts, and insofar as DC markets are being closed to LDC exports (see Kaplinsky, 1981), then the spectacular export-led growth performance of some LDCs in the 1970s may come under severe threat in the coming decade.

The inescapable overall policy conclusion must be, therefore, that if LDCs intend to maintain or increase their penetration of DC markets for manufactures, they have no option but to introduce CAD and similar technologies as rapidly as possible. But they are a number of factors that complicate this policy; unless there are considered carefully the attempts that LDC governments and firms make to utilize these technologies will not realize the desired ends. Four major caveats of this sort emerge with regard to CAD which are probably generalizable to other electronics-related technologies:

(a) *The appropriateness* [4] *of the technology*. As we saw earlier the introduction of CAD changes design from a labour-intensive activity (with a cost per workplace of less than $2,000) to a significantly more capital-intensive one (frequently involving capital costs per workplace of over $80,000). LDCs who are faced with capital constraints, foreign exchange constraints (for CAD systems will inevitably have to be imported) and high levels of unemployment will consequently find themselves using what appears to be a seemingly inappropriate technology but which, by virtue of its performance, renders manual design techniques sub-optimal. This is not the first time that LDCs have been faced with the trade-off between technological efficiency and appropriateness; nor, is the dilemma confined to LDCs, since many DCs currently face balance of payments constraints and untypically high levels of unemployment. So despite the undesirable characteristics of such technologies, if LDC producers (and DC producers for that matter) wish to maintain their competitiveness in world markets they will be forced to make more active use of CAD-type technologies, however inappropriate they might seem.

(b) *Skill requirements*. It can be seen from the discussion in Chapter 8 that the skills required to operate CAD systems are almost certainly lower than those existing in the traditional drawing and design offices. From this point of view, therefore, there are no grounds for arguing that LDC firms will be unable to utilize the new technology. However in this earlier analysis it emerged that the implications for management are more fundamental than those for operators. Not only do managers have to ensure that the design and draughting offices are organized in different ways, and to be aware of the systems-gains that CAD technology permits, but management

also has to possess the power to implement the changes that are required. In fact there are many reasons to suppose that it is easier to implement systems re-organization in LDCs than in DCs. In part this is because of the differing relative powers of labour and capital in the two environments. But it is also conceivable that such systems will be easily introduced in greenfield sites in LDCs, whereas in DCs existing enterprises will struggle with established procedures and interest groups in changing existing work practices. Clearly, in the latter case, the power of labour to inhibit the efficient introduction of electronics technologies (especially when the survival of the whole categories of activity are concerned) [5] will be significant. With respect to the ability of LDC management to comprehend the required changes we must keep an open mind since we are uncertain from the limited sample of users visited to what extent this is a function of prior experience, or of training. To the extent that it is the former, then we may well anticipate differential bottlenecks in the ability of LDC and DC producers to use the new technology.
(c) *Synergy and external economies*. At various points in this study we have observed that the major gains from the utilization of CAD technology probably lie in the capturing of systems gains as we move towards the 'factory of the future'. By definition these external economies require the complementary utilization of electronic techniques in information control and production/distribution. Thus, to the extent that LDC producers have not been faced by the high wage costs that have forced DC enterprises to introduce these complementary downstream technologies, then it is unlikely that they will be able to capture these significant external economies. How significant a problem this is will depend upon the particular process involved in each sector of activity. Moreover to some extent these downstream systems gains are reflected in cost of production rather than in product characteristics, lead time, marketing and control over technology. Low wage costs in LDCs may then continue to allow for the profitable use of a particular electronics technology such as CAD is restricted to the design/draughting sub-process and the downstream and upstream benefits are consequently lost.
(d) *Distance from suppliers and other users*. We observed in Chapters 3, 5, 6 and 8 that the immaturity of particular CAD applications programs usually leads to a hand-holding relationship between suppliers and users, as well as cross fertilization with other users. As the major mode of competition moves in the 1980s from the strength of specific applications software to price and servicing, so most CAD suppliers are offering users the capability to respond

to their problems within four to eight hours. From the users point of view this rapid response is essential not only because the high capital cost of CAD requires high utilization rates, but also because their reliance on a centralized data base makes them highly dependent on the unhindered functioning of their CAD systems. From the suppliers point of view this rapid response can only be provided if service centres are distributed widely through their user base. Although these points are specific to CAD, we can anticipate that other software-intensive technologies will require a similarly close relationship between suppliers and users. We have seen a number of cases where the sale of CAD systems to LDCs has either led to the effective severance of the relationship between suppliers and users, or the continuation of the relationship at a very high cost. In the case of Brazil, where the ratio of service plus applications engineers to CAD systems was 1:1 rather than between 1:4 and 1:8 in the United States, the costs were largely borne by the CAD supplier which anticipated a growing number of users in the expanding automobile industry. But in another case of a petroleum TNC, which purchased three CAD systems for mapping, the extra costs were borne by the user. More generally, though, LDC users were left on their own. On the basis of the experience of DC users visited, this isolation must be seen as a disadvantage despite the possibility that isolated users will be forced to learn more about the CAD software itself than their more pampered counterparts. The disadvantages are probably most apparent in relation to the extension of the CAD systems to wider uses outside of the particular applications programs for which the systems were initially purchased, whatever the familiarity which the isolated user has perforce had to develop with any particular set of applications programs. Given that the most effective United States and United Kingdom users had achieved their major design gains by widening applications in this way, this restriction in use is likely to be a major disability.

We can sum up the situation in the following way. The imperative for LDC producers (who wish to maintain or increase their competitiveness in DC markets) to utilize electronics-related technologies such as CAD is very strong, although it will clearly vary by sector and over time. While there are probably no absolute barriers to their effective introduction in LDCs, there are a number of factors which lead us to believe that LDCs face a number of disabilities in using CAD and other electronics technologies.

At a higher level of abstraction (e.g., within the framework of neo-classical economic theory) some would conclude that policy intervention is unnecessary – market forces will lead to the rapid diffusion of CAD type technologies to LDCs. But the evidence of this sectoral study suggests that the existing market structure is an inefficient transferer of information and that LDCs face some obstacles in using these technologies. Moreover, the oligopolistic CAD market structure, in which the competitive pressures which might lead to the active marketing of the technology in LDCs are dulled by the very rapidly growing market, are unlikely to speed up the diffusion of these technologies to LDCs. Therefore, we are forced to evaluate other methods of increasing the rate of diffusion, and two come to mind – TNCs using these technologies and LDC government intervention.

9.2 TNCs, CAD technology and the location of production

Since between 30 and 40 per cent of total world trade in manufacturing occurs within TNCs, and a substantial proportion of the rest also involves TNCs (either as buyers and sellers, or as buyers or sellers), their locational decisions in the context of changing technological, economic and political environments, are of critical importance.

In Chapter 5 we noted that during most of the 1970s the market structure of the CAD supplying industry had been dominated by new, specialized firms. By the end of the decade three new forms of concentration were becoming apparent in the CAD supplying sector (see Kaplinsky, forthcoming). First, in addition to IBM, a number of existing electronics firms producing computers (mainframe and mini) and terminal screens were beginning to penetrate the industry. Second, there has been an increasing tendency for established engineering firms to encapsulate CAD technology, in their move towards the supply of automated technologies to industry. And third, we are seeing a trend towards the multinationalization of production of CAD technology itself – Computervision, the market leader, is currently assessing the feasibility of manufacturing CAD systems somewhere in Europe. Thus, despite the recent trend towards new small firms supplying dedicated microprocessor systems, at an aggregate level we see a tendency towards the increasing incorporation of CAD technology within TNCs.

But more importantly it appears as though there is also a trend towards TNCs being particularly heavy users of the technology. Unlike the early 1970s when CAD supplying firms operated at a low scale and sold to any knowledgeable firm that was able to use CAD (and many of these were high-technology, national firms), the growing scale of the market is forcing CAD suppliers to take TNCs, who are potentially purchasers of multiple systems,[6] as their prime target. One CAD supplier described itself as following the DEC strategy of working its way down the *Fortune 500* list of United States firms.

So TNCs are not only heavily represented as manufacturers in LDCs, but also as suppliers and users of the new CAD technology. How will they then react to the new technological environment in their future locational decisions? There are two alternative scenarios here. The first is that the advantages of CAD and other electronics-related technologies reduce the incentive to site production in LDCs, since they not only diminish the comparative advantage of low cost labour in LDCs, but also by locating in LDCs, TNCs will suffer the disadvantage of distance from final markets and technology supplying firms. The alternative view is equally tenable – for example, forthcoming developments in communications technologies and in distributed CAD technology will make it feasible for TNCs to locate design in advanced countries and by direct electronic transmission of design parameters, to maintain productive facilities in LDCs. Moreover, the weaker power of labour in LDCs will make it easier to establish greenfield sites which allow for the capturing of systems gains, as the move to the automated factory quickens. In this way by locating design in DCs and production in LDCs it will be possible for TNCs to get the best of both worlds.

It is unfortunately premature to judge which of these two alternative paths TNCs will take. In part we just do not know enough about the impact of electronics in a number of other related sectors. But, in part, it also depends upon the move towards protectionism in DCs which is likely to emerge as the long-wave cycle continues its downswing and labour-saving electronics innovations diffuse more widely through these economies. But clearly, their response to these changing economic, political and technological parameters will be of crucial significance to the international location of production.

9.3 Policy implications for LDC governments

So far, the market process has not in itself led to the widespread diffusion of CAD technology to LDCs. Nor. for reasons discussed above, can we expect it to do so in the immediate future, given the fact that almost all CAD suppliers are facing almost insurmountable difficulties in coping with the pace of market growth in the United States, Europe and Japan,[7] given the extra costs of their servicing LDC users, and given the imperfect flow of information with regard to the benefits derived from using CAD.

Indeed even DC governments recognize the limitations of the market in producing the incentive required to diffuse CAD technology as rapidly as competitive conditions require. For example, the British government heavily subsidises the CAD Centre in Cambridge which is designed to spread the dissemination of CAD to industry: it also provides aid to firms that pioneer the use of the technology. The German government provides tax incentives to firms using CAD (*Anderson Report*, Vol. 1, No. 4), while the Norwegian Government heavily aids CAD supplying firms. More recently the Canadian Government (Government of Canada, 1980) has established a special programme of action for the introduction of CAD technology that notes *inter alia*, that:

> Productivity will be especially important to the Canadian manufacturing industries in the 1980s if traditional markets are to be retained and new ones gained . . . .
>
> In this context, the rapidly emerging use of Computer Aided Design and Computer Aided Manufacturing (CAD/CAM) technology is of special importance.

Consequently LDC governments cannot remain passive if they are to stem a re-opening of the technological gap between DCs and LDCs and maintain or increase their share of DC markets. In addition to specific steps which will be necessary in particular industries (e.g., shipbuilding), a series of more wide-ranging policies will have to be implemented in a number of areas, namely:

(a) *Increasing awareness*. Local firms, state and privately owned, will need to be informed of the potential benefits associated with the use of CAD type technologies and of the difficulties and skills which are involved. It would appear as if the international flow of information in this respect is highly imperfect and that an active awareness-raising programme by governments is essential.

(b) *Training*. It will be necessary to specifically train nationals for the advent of these technologies. This is not so important at the level of operators, where by and large supplying firms provide the required services (albeit at a cost), but more importantly in relation to management. Of particular importance, here, is the awareness by management of the potential systems gains that automation and CAD permits. A second arena in which specific training is desirable (and which is currently undertaken in most DCs) is to build CAD and other electronics technologies into the syllabi of engineering and related courses.

(c) *Aid.* Just as many DC governments are doing, LDCs will inevitably have to aid the purchase of CAD and related technologies by indigenous firms. In most LDCs electronics equipment sells for between two and three times the price prevailing in DCs. This is due to a variety of reasons including the low level of sales (which does not allow suppliers to spread out unit overhead costs), tariffs and local sourcing policies.[8] Therefore, in some cases this government assistance will therefore have to involve financial subsidies and/or access to scarce foreign currency resources. This aid will not only speed up the purchase of these technologies by particular firms, but can also assist in the dissemination of information. For example, the experience of the British government has been that the provision of aid to certain key users has provided fairly open access of these systems to other firms in the United Kingdom who have been exploring the introduction of CAD technology.

(d) *Services centres, research centres and bureaux.* In DCs the establishment of service centres lies in the hands of the CAD supplying firms. However, as we have already seen, servicing LDC users is usually a high-cost operation and this is one of the reasons why CAD suppliers have been slow in marketing in LDCs. There are two major options open to LDC governments here. Either they can directly subsidize supplying firms in the establishment of service centres. Or, as an alternative, they can establish 'machine transparent service centres', that is service engineers and software specialists [9] who are sufficiently knowledgeable to service a variety of different systems. This latter path presents immediate difficulties in that some servicing problems (for example involving specific sets of 'firmware' hardware) are highly firm-specific. But on the other hand, it will allow LDCs to take advantage of a variety of different supplying firms' software strengths and will at the same time inevitably increase the extent of learning with regard to electronic technologies, particularly if these 'machine transparent' service centres are linked to universities and research centres. Clearly this latter course presents difficulties and its feasibility needs closer examination. Associated with this strategy of establishing service centres is the possibility of establishing bureaux to service a number of users and thus protect them from their ignorance and from the high entry costs.[10] Bureaux do exist in DCs although they suffer from the fact that most users develop highly specific systems skills. Nevertheless, they have a clear role to play in any country's national policy response.

(f) *Indigenous firms as producers of CAD systems.* As we saw earlier, given the nature of the origins of CAD technology, it is unlikely that LDCs will become viable producers of analogous CAD technologies in the foreseeable future. However, there is no reason why a particular LDC government (or perhaps a group of governments) should not buy up an existing CAD supplier in the United States or Europe. Purely as a speculative investment there may well be substantial gains in a judicious purchase of this sort (see Chapter 5), but there may be a number of other spin-offs that might result from such a bold step, including the greater likelihood that CAD technology will be actively marketed in the relevant LDCs.

9.4 In conclusion

In the first two sections of this study we argued that in the context of long-wave cycles of economic activity, there are reasons to suppose that the growing penetration of DCs by LDC exporters of manufacturers would be severely threatened in the 1980s as microelectronic-related innovations diffuse through the advanced economies. In particular we noted that there were signs that the technological gap between DCs and LDCs, which appears to have narrowed in the 1960s, and 1970s, might well re-open in the 1980s. Moveover, we also predicted that electronics-related innovations would lead to growing labour displacement in DCs and an increasing tendency towards import controls.

This study on CAD has been designed to explore the extent to which these phenomena have become apparent in a single sector. The conclusion, which has largely been summarized in the earlier part of this section, bear out our three major hypotheses – namely, that the diffusion of electronics provides very significant gains to user firms, that it is likely to lead to the displacement of a significant number of draughtspersons in firms with traditional design offices and largely stagnant markets and that the technology has barely penetrated LDCs and shows little sign of doing so in the immediate future, despite the absence of a significant skill barrier at the operational level. Moreover the primary impact of CAD and other electronics technologies appears to be in precisely those non-traditional sectors in which LDC exports have grown most rapidly and in which they hope to specialize in the coming decade.

The problem with generalizing (this case study) is that as we have seen earlier, there are to our knowledge almost no other sectoral studies that explore these issues. We are consequently largely in ignorance as to whether CAD is representative of a more general trend, or whether it is a single, atypical example. We conclude therefore with the observation that *if* the emergence and diffusion of other electronics technologies in downstream users during the rationalizing stage of the long-wave cycle proximates in any meaningful sense to the CAD sector, then there is an urgent need for LDC governments to act decisively to arrest what appears to be a dangerous trend in export competitiveness.

The major question must be – how representative is the CAD sector?

Notes

1 We refer here to the microeconomic impact. At the macroeconomic level it is probable that the labour displacing characteristics of the technology will restrict access into DC markets for LDC manufactures (see Kaplinsky, 1981).

2 That is, that they are sub-optimal at any combination of factor prices, due to the fact that they have lower productivities of both capital and labour than their economically superior counterparts.

3 For example the Ford Motor Corporation is pursuing a global 'AJ' (i.e., After Japan) strategy which involves the rapid introduction of better designed and produced cars. Various electronics technologies, including CAD, modellers and robots, are the key components of this AJ strategy.

4 By appropriateness we refer in this case to the relationship between a technology and the economy in which it is introduced. This is, of course, a very restricted view of the concept but is adequate for the purposes of the discussion which follows.

5 In one case in the aircraft industry, a CAD user observed that the productivity of the system was diminished by deliberate 'sabotage' by a draughtsperson whose job was directly at stake. In the United Kingdom there are a couple of notable examples of firms whose systems were hamstrung by disputes with the trade unions – especially the draughtspersons union (TASS) whose members were most threatened by its introduction. But, more significantly, it was the view of a number of CAD suppliers that in general British trade unions were amenable to the introduction of CAD and often used it as a bargaining point for higher wages – the more common view was that conservative British management was the major impediment to the wider diffusion of the technology.

6 By the end of 1978, for example, General Motors alone possessed more than 300 CAD terminals in the United States: by 1983 it was estimated that they would be using over 3,000 (*Anderson Report*, Vol. 1, No. 1, September 1978). General Electric had over 100 systems in use in 1980 (each of which comprises a number of terminals) and expected to purchase another twenty-five systems in 1981 alone.

7 As one CAD supplier reported, market shares in the 1980s will depend upon the extent to which CAD supplying firms can actually manage (i.e., provide the organizational facilities to cope with) compound growth rates of over 70 per cent.

8 Brazil and India both virtually prohibit the import of mini- and microcomputers in order to protect their indigenous electronics industries.

9 Some American firms use a number of different CAD systems which are managed centrally. Clearly, therefore, there are some levels at which 'machine transparent' software and organizational skills are interchangeable between different CAD systems.

10 CAD is seldom viable with less than ten design staff and for design it currently involves entry costs of at least $160,000 for a basic two-terminal system.

APPENDIX I: Research methodology

Interview technique

From inception to completion only six months was available for this case study to be undertaken. It was thought that this would be sufficient to undertake basic fieldwork and visits were made to a variety of supplying and using firms. Unfortunately the lack of time curtailed the number of users who could be visited (not only in continental Europe and Japan but also in LDCs), as well as a more detailed investigation of CAD firms specializing in software (especially in solid modelling).

Two alternative interview techniques are possible in such field visits. The first is that of a detailed questionnaire, and the second comprises of free-flowing but structured discussion, backed up by check lists to ensure that all major areas of interest are covered. The former technique was rejected since experience has shown that it pre-determines responses and antagonises the interviewees. Consequently, structured discursive interviews were held, based upon the following two checklists, one for CAD vendors and one for CAD users.

Checklist for CAD vendors

(*a*) *Background*
(i) Firm
(ii) Size and growth over time
    turnover
    employment
(iii) Product range
    hardware
    software
    other products
(iv) Sectoral distribution of sales
(v) Geographical distribution of sales
(vi) Marketing policy
    DC/LDC
    small/large firms
    packaged/unpackaged
    mode of competition

(c) *The market*
(i) Price trends of systems
(ii) Availability of systems – supply vs. demand
(iii) Market structures
globally
within country } past, present, and future
within sector
(iv) Hardware vs. software
(v) Packaging vs. unpackaging
(vi) Sales by sector, area over time
(vii) Barriers to entry

(d) *Proprietary rights*
(i) Specific or general skills
(ii) Property rights
(iii) Technical interchange
(iv) Other methods of appropriation

(e) *Economic benefits of CAD*
(i) When it is justified to use?
(ii) Savings in labour
in drawings
absence of drawings
and in capital
working
fixed
(iii) Materials
(iv) Scale implications of production
(v) On product
leadtime
quality: differentiation of existing and new products
(vi) Systems gains
(vii) Avoiding bottlenecks
(viii) Impact upon market share
(ix) Other

(f) *Skill implications in use*
(i) Is there a learning curve? } operator and management
(ii) What skills are required?

(g) *Diffusion*
(i) Projected rates
(ii) Which sectors?
(iii) Obstacles

(h) The future
(i) Changes in technology
(ii) Changes in price

(i) Contacts
Within CAD supplying industry
Successful users
Unsuccessful users
Users who compete with LDCs

Checklist for CAD Users

(a) Background
(i) Firm
(ii) Size and growth
    turnover
    employment
(iii) Products, market structure and market share
(iv) Breakdown of costs
    labour
    capital
    materials
    energy
(v) Design as a proportion of product costs

(b) CAD system used
(i) Hardware Fixed – historic and replacement cost
(ii) Software Variable – historic and replacement cost
(iii) How do they define CAD?
(iv) Their use
    draughting
    design
    downstream
    other
(v) Capacity utilization
(vi) Search procedure

(c) Manpower
(i) Did it lean on existing skills?
(ii) Training costs and programmes
(iii) Learning curves – manager and operator
(iv) Profile of skills – with CAD vs. without CAD

(d) Benefits

(i) Why was CAD introduced?
(ii) Were these achieved?
(iii) Additional benefits?
Note:
1 Product
Leadtime
Improvements in
appearance
operating characteristics
New products
2 Process
Labour
in drawing
absence of drawings
Capital
fixed
working
Materials
Flexibility capacity utilization
Scale
3 General
Avoid bottlenecks
Systems gains
e.g., in clerical organization of firm
4 Other
(iv) Impact upon market share

(e) Proprietary rights over own software

(i) Specific or general
(ii) Property rights
(iii) Sales of CAD software

(f) Technology

(i) What changes in hardware are desirable?
(ii) What changes in software are desirable?

Sample frame

All the major representatives of the United States turnkey firms operating in the United Kingdom were interviewed, some in the United Kingdom and others in the United States and the United Kingdom. In some cases multiple visits were made. A list of users was obtained from each of these vendors and visits were made to a significant proportion of each firm's customers, selected upon the basis of:

(a) Their sectoral activity (e.g., mechanical, civil or electronic engineering). While attempts were made to cover most major sectors, the emphasis was placed upon mechanical engineering since it is the sector in which most LDCs hope to compete in DC markets in the 1980s, and because it is probably the major growth area of CAD use over the coming few years.
(b) Particularly successful users.
(c) Particularly unsuccessful users.

Visits were made to a number of industry observers, particularly in the United States, and analysis was also based upon a variety of published documents.

In total twenty-four users were visited. Of these five were interviewed in the United States and the remainder in the United Kingdom. Six were only producing in their country of incorporation and the remainder were TNCs. Turnover varied and ranged from $1.7 million (a CAD bureau) to $10 million (the smallest manufacturers) to billions of dollars. The firm with the smallest design staff provided two terminals for three designers. Of the twenty-four users, two were CAD bureaux, three used it predominantly for electronics, six predominantly for civil/structural/process engineering and mapping, and the remainder predominantly for mechanical.

Thirteen CAD suppliers were interviewed with a turnover ranging from zero to $26 billion. Of these only one was a wholly British firm, the rest were American. Additional information on these and other firms were obtained from the *Anderson Report*, the *Harvard Newsletter on Computer Graphics* and the various Merril Lynch reports.

BIBLIOGRAPHY

Advisory Council for Applied Research and Development (1980) *Computer-Aided Design and Manufacture*, London, HMSO.

Arnold, E. (1981) 'The manpower implications of computer aided design in the UK engineering industry', *Proceedings of the British Computer Society Conference*, July 1981.

Arnold, E. and Senker, P. (1982) *Designing the Future: A Report for the Engineering Industry Training Board*, Brighton, Science Policy Research Unit.

Arrow, K. J. (1962) 'Economic welfare and the allocation of resources for invention', in *The Rate and Diffusion of Inventive Activity: Economic and Social Factors*, National Bureau of Economic Research, Princetown University Press.

Barron, I. and Curnow, R. (1979) *The Future with Information Technology*, London, Frances Pinter.

Bell, D. (1974) *The Coming of the Post Industrial Society: A Venture in Social Forecasting*, New York, Basic Books.

Bergsman, J. (1979) 'Growth and equity in semi-industrialised countries', *World Bank Staff Working Paper*, No. 351.

Besant, C. B. (1980) *Computer-Aided Design and Manufacture*, Chichester, Ellis Horwood Ltd.

Blyth Eastman Paine Webber Inc. and Alex Brown and Sons (1980) Prospectus for issue of 1,000,000 shares of Common Stock of Applicon, 20 July 1980.

Bosch, L. Lang Lendorff, G., Rothenberg, P. and Slebzer, V. (1978) *CAD-Berichte: The CAD Support Project*, Kernforschungszentrum Karlsruhe.

Braun, E. and MacDonald, S. (1978) *Revolution in Miniature: The History and Impact of Semiconductor Electronics*, Cambridge, Cambridge University Press.

CAD Centre (1978) *Computer Aided Design: An Appraisal of the Present State of the Art*, Cambridge.

Calma, (1978) *Interactive CAD: Considerations for Architects, Engineers and Constructors*, Sunnyvale, Calma.

Chenery, H. and Keesing, D. B. (1978) 'The changing role and composition of LDC exports', *Symposium on the Past and Prospects of the Economic World Order*, Institute of International Economic Studies, August.

Clark, J. Freeman, C. and Soete, L. (1980) 'Long waves and Technological developments in the 20th century' (mimeo), Brighton.

Computervision (1980) *The CAD/CAM Handbook*, Bedford, Mass.

Cooley, M. (1980) 'The designer in the 1980s – the deskiller deskilled', *Design Studies*, Vol. 1, No. 4.

Dean Winter Reynolds Inc. and The Robinson-Humphrey Company Inc. (1981) Preliminary Prospectus for Issue of 1,500,000 Shares of Common Stock of Intergraph Corporation, 25 February, 1981.

Department of Industry (1978) *Computer Aided Design and Manufacture: Survey of Available Systems and Scope for their Application in Mechanical*

Engineering, London, Mechanical Engineering and Machine Tools Requirements Board HMSO.

Drexel Burnham Lambert Inc., (1981) Preliminary Prospectus for issue of 600 shares of common stock of Gerber Systems Technology Inc., 26 February 1981.

English, J. M. (ed.) (1968) *Cost Effectiveness: The Economic Evaluation of Engineered Systems*, New York, Wiley.

Fidgett, T. (1969) 'A summary of information on the employment, training and supply of technician engineers and technicians in the engineering industry', *Engineering Industry Training Board Reference Paper*, RP/3/79, London.

Forester, T. (1978) 'The microelectronics revolution', *New Society*, 9 November.

Frazier, M. (1981) 'Free trade zones: a growing phenomenon around the globe', *Transatlantic Perspectives*, No. 4, January.

Freeman, C. (1977) 'The Kondratieff long waves, technical change and unemployment', *Proceedings of OECD Meeting of Experts on Structural Determinants of Employment and Unemployment,* Paris.

Freeman, C. (1978) 'Technical change and unemployment', *Paper prepared for 6 County Project Meetings in Paris*, Brighton, Science Policy Research Unit.

Freeman, C. (1979) 'Technical change and unemployment' in Encel, S. and Ronayne, J. (ed.) *Science, Technology and International Perspectatives*, Oxford, Pergamon.

Freeman, C. (1981) 'Economic implications of microelectronics', in Cohen, D. (ed.) *Agenda for Britain I: Micro Policy (Choices for the Eighties)*, London, Philip Allan.

Government of Canada (1980) *Strategy for Survival: Issues and Recommendations Concerning the Implementation and Impact of CAD/CAM Technology in Canadian Industry*, A report by the CAD/CAM Technology Advancement Council, Dept. of Industry, Trade and Commerce, Ottawa, September.

Hambrecht and Quist (1979) Prospectus for issue of 500,000 shares of Common Stock of Auto-trol Technology Corporation, San Francisco, 24 January 1979.

Henwood, T. D. (1980) *Computer Graphics Industry: In Its Early Growth Stages*, New York, First Boston Research.

Hill, T. P. (1979) *Profits and Rates of Return*, Paris, OECD.

Hoffman, K. and Rush, H. (1980) 'Microelectronics, industry, and the Third World', *Futures*, August, pp. 289-302.

Hoffman, K. and Rush, H. (1981) 'Apparel and microelectronics: the impact of technical change on a global industry', (mimeo) Brighton, Science Policy Research Unit.

Jacobs, G. (1980) Designing for improved value', *Engineering*, February, pp. 178-182.

Kaplinsky, R. (1981) 'Radical technical change and export-oriented industrialisation: the impact of microelectronics', *Vierteljahresberichte*, No. 83.

Kaplinsky, R. (ed.) *Comparative Advantage in an Automating World*, IDS Bulletin, Vol. 13, No. 2, March 1982.

Kaplinsky, R. (1983) *Automation in a Crisis*, London, Pluto Press.

Kaplinsky, R. (forthcoming), 'Firm Size and Technical Change Reconsidered', *Journal of Industrial Economics.*

Katz, J. (1978) *Technological Change, Economic Development and Intra and Extra Regional Relations in Latin America*, IDB/ECLA Research Programme in Science and Technology, Working Paper No. 30, October.

Kurlack, T. (1980) *Computer Aided Design and Manufacturing Industry CAD/CAM*, New York, Merrill Lynch.
Kurlack, T. (1981) *CAD/CAM Review and Outlook*, New York, Merrill Lynch.
Kleinknecht, A. (1980) 'Considerations on the renaissance of the 'long-waves' in economic life [Kondratieff Cycles] ', (mimeo).
Kline, M. B. and Lifson, M. W. (1968) 'Systems engineering' in *English*, 1968.
Knickerbocker, F. T. (1973) *Oligopolistic Reaction and Multinational Enterprise*, Boston, Harvard Business School.
Kraft, P. (1977) *Programmers and Managers: The Routinization of Computer Programmers in the US*, New York, Springer-Verlag.
Lall, S. (1979) *Developing Countries as Exporters of Technology and Capital Goods: The Indian Experience*, Oxford (mimeo).
Laurie, P. (1980) *The Micro Revolution*, London, Futura.
Mandel, E. (1980) *The Second Slump: A Marxist Analysis of Recession in the Seventies*, London, New Left Review Books.
Nora, S. and Minc, A. (1978) *'L'Informatisation de la Societe'*, Paris, La Documentations Francais. Published in English as *The Computerization of Society*, Cambridge, Mass. 1980 MIT Press.
Noyce, R. (1977) 'Microelectronics', *Scientific American*, September.
O'Brien, P. (1981) 'Third World industrial enterprises as exporters of technology – recent trends and underlying causes', *Vierteljahresberichte*, No. 83, March.
OECD (1979) *Interfutures: Research Project on the Future Development of Advanced Industrial Societies in Harmony with that of Developing Countries*, Final Report, Paris, OECD.
Plesch, P. A. (1978) *Developing Countries Exports of Electronics and Electrical Engineering Products*, IBRD Economics of Industry Division, Development Economics Dept., Washington.
Rada, J. F. (1979) 'Microelectronics, information, technology and its effects on developing countries', paper prepared for the Conference on Socio-economic Problems and Potentialities of the Application of Microelectronics at Work, The Netherlands, 19–24 September 1979.
Rader, M. and Wingert, B. (1981) *Computer Aided Design in Great Britain and the Federal Republic of Germany – Current Trends and Impacts*, Karlsruhe, Kerforschungszentrum Karlsruhe GmbH.
Rostow, W. W. (1978) *The World Economy: History and Prospect*, Basingstoke, Macmillan.
Scriberras, E. (1979) 'Technology transfer to developing countries – implications for member countries', Science and Technology Policy, *OECD Television and Related Products Sector Final Report*, Paris.
Senker, P. (1980) *Microelectronics and the Engineering Industry: The Need for Skills*, London, Frances Pinter.
Sim, R. M. (1980) 'Computer aided manufacture in batch production', *Background Paper for Auew (TASS) Conference on Computer Technology and Employment on 16/9/1980*, East Kilbride, National Engineering Laboratories.
Soete, L. (1977) 'Size of form, oligopoly and research: a reappraisal', *Extrait de Reseaux*, Nos. 35–6.
Soete, L. (1979) 'Firm size and inventive activity', *European Economic Review*, 12, pp. 319–40.
Stewart, F. (1978) *Technology and Underdevelopment*, London, Macmillan.
The Anderson Report, Vol. 1, No. 1 to Vol. 3, No. 8, Simi Valley, Cal., Ander-

son Publishing Company.
The Harvard Newsletter on Computer Graphics, Vol. 1, No. 1 to Vol 3, No. 10, Harvard Laboratory for Computer Graphics, Sudbury, Mass.
Transatlantic Perspectives (1981) 'The Western Automobile Industry at the Crossroads', *Transatlantic Perspectives*, No. 4, January.
UNIDO (1979) *Industry 2000: New Perspectives*, New York.
UNIDO (1980) *World Industry Since 1960: Progress and Prospects*, New York.
United States Central Intelligence Agency (1980) *Developed Country Imports of Manufactures from LDCs*, ER-80-10476, Washington.
United States Department of Labour (1969) *Tomorrow's Manpower Needs*, Dept. of Labour Statistics Bulletin No. 1606.
Vernon, R. (1966) 'International investment and international trade in the product cycle', *Quarterly Journal of Economics*, Vol. 80, No. 2, pp. 190–207.

GLOSSARY

Entries in quotation marks are drawn from a more specialized glossary published in Computervision (1980).

ACE Architecture, construction and civil engineering, a term used to describe a particular set of markets for CAD equipment.
Add-on memory The capability to store information outside of the central computer.
Applications software CAD software programs which have been developed to meet the requirements of particular users.
Batch processing 'A group of jobs to be run on a computer in succession, without human intervention'. Turnaround of work is consequently not instantaneous and the process is consequently not interactive.
Bills of materials 'A listing of all the sub-assemblies parts, materials and quantities required to make one assembled product'.
Binary code 'System of representing numbers [or sub-sets of logic] using the characters 0 and 1 to represent any number'.
Bit A single binary code.
CAD (computer-aided design) The use of computers in design with interactive, graphic capabilities.
CAM (computer-aided manufacture) The use of 'computers technology to manage and control the operations of manufacturing facility'.
Chips See integrated circuit.
Component 'Symbol which has a physical meaning e.g., switch, resistor, capacitor'.
Core memory 'Main memory resident in the computer'.
Data base 'Comprehensive collection of information having predetermined structure and organisation suitable for communication, interpretation, or processing.
Dedicated A set of hardware or software confined to a single use.
Digital A synonym for a system working with binary code logic.
Digitizer A method of converting drawings into digital numerical co-ordinates that can be processed by a computer.
Disc One method of storing information outside of the central computers. Discs can be either rigid or pliable ('floppy discs') and require a disc drive (an equivalent to a record player) for use.
Downstream Activities in production control or manufacture/distribution affected by decisions made in design.
Finite element analysis A system of 'wire-frame' meshes used to 'determine the structural integrity of a mechanical path by mathematical simulation'.
Firmware Software instructions written into hardware.
Graphics software CAD software enabling the user to draw lines, arcs, circles etc. and to manipulate these on the terminal screen.
Hardware The physical elements of electronics technology.
Interactive The capability that allows the CAD user to manipulate and modify information and software in a computer system without delay.
Intergrated circuit 'Tiny complex of electronic components and their connections produced on a slice of material such as silicon'.

Large scale intergration (LSI) Intergrated circuits, each with the capacity to process more than 32,000 bits of information but less than about 262,000 bits of information.
Light pen A pen-like object that allows the operator to write on the screen.
Machine code The basic binary code used by all computers.
Magnetic tape Similar to reel-type tape recorders, used for the storage of digital information outside of the central computer.
Mainframe Large computers.
Menus 'Input device consisting of command squares on a digitizing surface. It eliminates the need for an input keyboard for common commands'.
Memory Storage of digital information. See, also, Add-on, Core, Disc, Magnetic Tape.
Microcomputer The smallest-sized computer, usually based upon a single microprocessor.
Microprocessor A single intergrated circuit with logical capability.
Minicomputer Intermediate sized computers.
Nesting The method of arranging the patterns of parts to make optimal use of materials from which they are to be cut.
Numerical control A control system based upon digital logic.
Numerical co-ordinates Information represented in binary code.
Operating system Basic software that determines the functioning of hardware.
Paper tape Like a ticker tape on Telex, the major method of feeding digital information into numerically controlled machinery.
Parts lists See 'Bills of materials'.
Parts programming The process whereby the optimum path is defined for cutting parts on a machine tool.
Plotter A method for representing graphical information, drawn from computerized information, on to paper or film.
Printed circuit boards (pcb) 'Insulated substrate (usually plastic upon which interconnected wiring [usually between components and integrated circuits] has been applied by photographic techniques'. Almost all electronic products (whether in TVs, computers or digital watches) are built around circuit boards.
Small-scale integration (SSI) Integrated circuits, each with the capacity to process less than 32,000 bits of information.
Software The instructions that enable electronic hardware to operate. See, also, 'Applications software', 'Graphic software', and 'Operating systems software'.
Storage See 'Memory'.
Terminal A television-like screen which allows CAD users to work graphically.
Turnkey CAD systems sold with a combined package of hardware and software.
Uncommitted logic array (ULA) Often called 'gate-arrays'. A method in which all the logic elements (called gates) can be built into the lower layers of an integrated circuit allowing the final user to make the connection between the different gates, each of which represents a bit of information. This is a rapidly evolving technology.
Vacuum tubes 'Valves' used in the days before the electronics industry produced 'solid state' transistors, integrated circuits, microprocessors and components.
Very large scale integration (VLSI) Integrated circuits, each with the capacity to process more than about 262,000 bits of information.
Wire frame The mask used in 'Finite element analysis'.
Workstation See 'Terminal'.

INDEX